Science, Reason, Modernity

forms of living

Stefanos Geroulanos and Todd Meyers, *series editors*

Science, Reason, Modernity

Readings for an Anthropology of the Contemporary

Edited by Anthony Stavrianakis,
Gaymon Bennett, and Lyle Fearnley

FORDHAM UNIVERSITY PRESS
NEW YORK 2015

Library of Congress Cataloging-in-Publication Data available
online at catalog.loc.gov.

Printed in the United States of America
17 16 15   5 4 3 2 1
First edition

*To Paul, for connecting us to the recent past
and opening up a near future*

CONTENTS

ACKNOWLEDGMENTS

This volume would not have been possible without the intellectual friendship and constant generosity of Paul Rabinow and the other members of Anthropological Research on the Contemporary (ARC). We would like to thank James Faubion for his magnanimous counsel and Todd Meyers and Stefanos Geroulanos for their support for this project and for giving it an intellectual home within the Forms of Living series. At Fordham we would like to thank Tom Lay for his care and guidance as well as to add our voices to others who have expressed profound gratitude for having had the opportunity to work, if only too briefly, with Helen Tartar.

Science, Reason, Modernity

Introduction: Contemporary Equipment
for Anthropological Problems of Modern Sciences

Anthony Stavrianakis, Gaymon Bennett, and Lyle Fearnley

Over the past four decades, anthropological, sociological, and historical studies of the modern sciences have made prominent contributions to the conceptual and methodological renovation of contemporary thought. Although these studies of science are united by a shared domain of inquiry, their growth has been accompanied by a rapid proliferation and diversification of concepts and methods. The student beginning a program of research into the sciences today is likely to be confronted by diffuse, if not contradictory, possible approaches toward her object of study. This reader does not provide an overview of these widely varying works and recent methodological innovations, nor does it attempt a total theoretical synthesis. Instead, here we provide an introduction to a longer legacy of philosophical and social scientific thinking about sciences and their integral role in shaping modernities, a legacy that has contributed to a specifically anthropological form of inquiry. Anthropological, in this case, refers not only to the institutional boundaries of an academic discipline but also to a mode of conceptualizing and addressing a problem: How do we analyze and diagnose the modern sciences in their troubled relationships with lived realities? Such an approach addresses the sciences as forms of life and illuminates how the diverse modes of reason, action, and passion that characterize the scientific life continue to shape our existences as late moderns.

The essays we have included in this book are drawn from this philosophical and social scientific legacy and mark out a pathway through it. Our hope is that by following this pathway, students and scholars working on sciences will be better equipped to think about scientific practices as anthropological problems. There are already several widely used readers in science and technology studies (STS).[1] The limitation of these existing readers, for our purposes here, lies in what one might call their pedagogical logic.

While these available resources give a necessary view on what has happened recently in the domain of STS, they fail to highlight the longer durational efforts to reflect on problems of science, medicine, and technology. They therefore do not sufficiently equip students with a genealogical attention to how science, medicine, and technology have figured prominently as objects of, and catalysts for, anthropological reflection (understood broadly and not in a narrow disciplinary sense), and thus how they have been problematized over the past century as defining features of modernity. We use the term "genealogical" here in the sense proposed by the historian and philosopher Michel Foucault: a selective history whose purpose is to provide the background needed to understand a contemporary problem—in this case the problem of thinking through and inquiring into problems of science as contemporary forms of life. We likewise use the term "problematized" in Foucault's sense: an attempt to specify and reflect on the historical dynamics by way of which problems of science and modernity become matters of serious reflection, as well as the multiple attempts to provide solutions to those problems.[2]

In short, the materials provided in this book have been arranged as a genealogical pathway across philosophical and social scientific works in which science has been problematized in relation to the breakdowns, limitations, and possibilities of modernity. These works have proven crucial, we think, for many contemporary anthropologies of science. Our aim is to make these works visible and to share what we think they can do for anthropological inquiry. If our composition is genealogical, our aim is therefore pedagogical.

We have ordered the materials in this book as an answer to the question of which authors and works we think students of anthropology ought to read in order to be better equipped to conduct inquiry into sciences and modernities today.

Some Genealogical Elements for Anthropologies of Modernity

HETERO-LOGOI: ANTHROPOLOGY AND MODERNITY

In his book *Anthropos Today*, the anthropologist Paul Rabinow proposes that *anthrōpos*—"(the) human (being),"—"is that being who suffers from too many *logoi*." That is to say, the human today is a being who suffers a heterogeneous plurality of reasoned discourses about its being.[3] For anthropologists of the contemporary world, this discursive proliferation has created a practical blockage: there is no longer any settled means—no set of settled terms, theories, or methods—for adjudicating or reconciling discordant claims about *anthrōpos* in its modernity. This is not for lack of trying. Twentieth-century anthropologists worked to overcome this plurality by connecting, opposing, and eliding humanity's *hetero-logoi* through of a series of seemingly comprehensive and stable terms, such as Man, Culture, Nature, and Society. And though these terms are in disrepair today, it is worth keeping in mind that they have been replaced by other equally comprehensive terms, such as globalization, the environment, networks, and neoliberalism. This new conceptual repertoire, like the old, is mobilized as a means of producing the sense that all of the discursive and practical cacophony in the world today somehow adds up. Rabinow is not the first to suggest that such efforts at theoretical unification, however tacit, are misplaced. He will not be the last to extol anthropologists to take seriously the reality of humanity's plurality of reasoned discourses and to refashion their tools, practices, and dispositions accordingly. As the historian of ideas Hans Blumenberg put it: with modernity, "Nature" can no longer be relied on to show "Man" the fixed limits of what "he" can and should make of "Himself."[4] Nor, we would add, can the unruly variety of Cultures or Societies be thought of as just so many expressions of the Human story awaiting the ethnographer's translation—as the anthropologist Clifford Geertz once suggested.[5] Attention to the *hetero-logoi* of modernity has become a diagnostic minimum for the practice of anthropological inquiry today.

Anthropologists over the last three decades have responded to the loss of Culture, Nature, and Society as stable points of reference by rethinking both their object of study and subject position. Rather than a classical anthropological attempt to "make the exotic Other familiar," anthropologists undertook efforts to "make the familiar strange," or, as it was sometimes put, to "anthropologize the West." Anthropologists working on modern sciences have been central to these multiple efforts, by critically engaging the categories and figures through which the "West" and the "moderns" have been imagined, as well as the material conditions of anthropological practice.[6] In doing so, as we argue in the next section, anthropologists of science have joined a broader movement toward investigating the rationalities of the moderns.[7] All these efforts have not only required shifts in research sites or topics; they have also required rethinking and redesigning anthropological concepts, modes of inquiry, and vocational sensibilities.

THE ANTHROPOLOGY OF MODERNITIES

From the 1960s, anthropologists in the United States, the United Kingdom, and France (among other places and often from other places) undertook anthropologies of European and American colonialism, imperialism, and the subordination and social oppression of women. This work was a critical response to changes in the world and in the institutional conditions of anthropological practice, ramifying developments in the academy and the world after the Second World War.[8] During the war, in the United States, many anthropologists had begun working in government service, alongside historians, social psychologists, political scientists, linguists, and sociologists. As the anthropologist Clifford Geertz explains:

> what had been an obscure, isolate, even reclusive, lone-wolf sort of discipline, concerned mainly with tribal ethnography, racial and linguistic classification, cultural evolution, and prehistory, changed in the course of a decade into the very model of a modern, policy conscious, corporate social-science.[9]

In the two decades after the war, and with a view to the shifting global conditions created by the Cold War and decolonization, the U.S. government continued to fund research consistent with its national interests. It gave priority to research that aimed at making the social sciences more problem centered, collaborative, cumulative, and explicitly linked to projects of modernization and international development. Support for the creation of the Department of Social Relations at Harvard in 1946, the Human Relations Area Files at Yale, and later, in 1960, the Committee on New Nations at Chicago were exemplary. At Harvard's Department of Social Relations, for example, both Geertz and the sociologist Robert Bellah conducted their thesis projects as part of programmatic efforts to generate an interconnected and comparative assessment of human affairs,[10] an assessment guided by the aim of understanding how economic modernization and rationalization were inflected by cultural and religious particularity.[11]

By the 1960s, many anthropologists who had participated in protest movements opposed to the war in Vietnam and who had witnessed firsthand the failings of postcolonialism and international development began to rethink critically the positions, practices, and premises that had come to dominate anthropological inquiry.[12] They pointed out that since the 1950s most people in the world—not least those caught in the binds of the Cold War and decolonization—were either not experiencing the progressive advance of rationalization and modernization or were experiencing it primarily as a source of exploitation and domination. The authors assembled in the volume edited by Dell Hymes, *Reinventing Anthropology* (1972), crystallized a series of political and methodological problems that had been articulated over the preceding decade. These include long enduring questions of responsibilities in ethnography, such as the place of academic elites in structures and institutions of power, as well as the challenge of how anthropologists might observe and participate in political transformations such as youth politics and the politics of race.[13]

THE POLITICS OF TRUTH: MENTALITIES AND REPRESENTATIONS

The formulation of projects critical of modernization catalyzed new intellectual and political approaches in anthropology: feminist intellectual projects posing the question of how to interface Euro-American social critiques of patriarchy and questions of the relations of sex and gender with ethnographic studies, the study of elites, and studies exploring the interconnection of anthropology and colonial governmental practice.[14] Through these projects anthropologists brought "the West" and "modernity" into sharper analytic focus. Among these critical assessments, two thematic strands ex-emplify the politics of truth that characterized this intensive and inventive period in anthropology: inquiry into the plural forms through which modernity was being experienced as a way of thought and life—historical modes for an anthropological analysis of "mentalities," as it has been called, and the analysis of the modes and forms of representation that had come to the fore in anthropology by the 1980s.

It is worth recalling that an anthropological sensibility toward historical experiences of modernity had already emerged in France, from the late 1920s. The historians Marc Bloch and Lucian Febvre conducted a series of histori-cal anthropologies of modes of experience in the early decades of the century. Later historians, such as Robert Mandrou, Georges Duby, and others, would subsequently locate these works under the banner of *l'histoire des men-talités*, or the "social history of modern mentalities."[15] Drawing upon Lucien Lévy-Bruhl's concept of "mentality," histories of modern mentalities were formulated in view of the perceived limits of work in the history of ideas and economic history, neither of which were paying sufficient attention to the interconnections of form and expression in quotidian life. Lévy-Bruhl had pioneered efforts to assess the *manner* of "primitive thought" rather than the *content* of "primitive beliefs."[16] He argued that primitive thought should not be studied against the logic of the "European individual" but rather, following Emile Durkheim's emphasis on "collective representations," in terms of the internal relations of its constitutive elements and in relation to the institutions of the social group.[17] By extending this mode of inquiry to the history of Eu-ropean mentalities, Bloch, Febvre, and others demonstrated that multiple his-torical forms and expressions of everyday thought coexist at a given moment in any given person. This European "person," these historians showed, is far from an anthropological constant.[18] Historians working in this vein studied core aspects of everyday European life, such as faith and death, tracing long-term structures and transformations as well as offering microhistorical ac-counts of these forms of European life from the middle ages to modernity.

While synchronistic tendencies in anthropology may endure, and while the external constraints for funding and writing within academic anthropol-ogy may encourage a presentist narrowness of view, nevertheless, the historicity of the experiences of becoming moderns, in heterogeneous forms, has only grown in thematic importance. We are all peoples with histories, as the Marxist anthropologist Eric Wolf described, and the anthropologist especially so.[19] Hence even if the people in question are not particularly historically or textually inclined,[20] the second chapter of the doctoral dissertation or published thesis, these days, is, almost obligatorily, "the history chapter," as the anthropologist George Marcus recently observed.[21]

There have of course been many serious efforts to experiment between and among history and anthropology.[22] Bernard S. Cohn's ethnohistorical work, for example, explicitly distinguished itself from previous efforts to understand the relation of colonialism to sociohistorical processes of rationalization and modernization. Cohn set out to develop an ethnohistorical anthropology of forms of knowledge and their formative effects. Trained at the same time as Geertz, and his colleague at Chicago, Cohn paid close attention to the disruptive rather than developmental features of decolonization as well as to what the study of such disruptions indicated *about* anthropological knowledge and prospects *for* anthropological knowledge. This linking of the object of anthropological observation with anthropological self-criticism entailed a reworking of core anthropological concepts: what is a village, a tribe, or a caste under conditions of colonial rule, and how does colonialism affect the historical transformation of relations between power and knowledge? Cohn's students, including Nicholas Dirks and Arjun Appadurai, would go on to develop their own critical trajectories and modes of engaging modernity as both an object of anthropological inquiry and occasion for rethinking anthropological practice.[23]

From 1980 forward further experimental innovations were introduced into the anthropological assessment of representations, including anthropology's own. The now-canonical *Writing Culture* (1986) marked a threshold moment in this regard.[24] The book brought together the diffuse countermovements to the Parsonsian anthropological mode of the 1950s and 1960s, using critical tools drawn from diverse sources of historical, literary, and philosophical work—including, and especially, work being done in France after the 1960s.[25] With respect to writing and representation, Marilyn Strathern, George Marcus, Michael Fischer, James Clifford, and others introduced new repertoires of concepts and narrative forms for anthropologists to use in rethinking the relation of field experience, textual production, and the problem of representation. By the end of the 1980s the question of representation and the question of power and modernity bound up in that question had become a major vector for the transformation of anthropological practice, breaking open previously unexamined relations among the ethics of fieldwork, the sites and situations of inquiry, and the politics of textual production.[26]

FROM MENTALITIES AND REPRESENTATIONS TO SCIENCES AND MODERNITIES

Contemporaneously with these critical experiments, a diverse set of anthropological and sociological attempts to understand the modern sciences began to take shape. These experiments were distinguished by the various modes by which they framed science as a problem of modernity and by the diverse forms of inquiry they invented or adjusted in order to carry out work on those forms.

A now-prominent form of inquiry emerged as part of the University of Edinburgh Science Studies Unit. The unit was established in 1966, initially directed by the radio astronomer David Edge, with the support of scientists such as the developmental biologist Conrad Hal Waddington. The purpose of the unit was to help bridge the growing divide between the "two cultures" of natural sciences and humanities, a divide diagnosed already in 1959 by the physicist and novelist C. P. Snow.[27] The Edinburgh School was shaped by the works of its leading figures, including Barry Barnes, David Bloor, Steven Shapin, and Donald Mackenzie, and it came to be associated with the manifestos of the "Strong Program" that declared that science must be treated with the same social scientific methods used to examine any other social practice.[28] The "Strong Program" emphasized science as "practice and culture,"[29] rejecting any essential division between science and society. This pioneering work would form a basis, among others, for later "laboratory studies," such as Bruno Latour and Steven Woolgar's account of the making of scientific facts in Roger Guillemin's laboratory's work on thyrotropin-releasing hormone.[30] Latour's methodological focus, with Michel Callon, at the Centre de Sociologie de l'Innovation of the École des Mines on material and semiotic relations of human and nonhuman assemblies was, furthermore, a fundamental step toward descriptive ontologies of actor-network theory.[31]

Feminist critiques of knowledge, nature, and the body provided another crucial form of anthropological inquiry into the sciences. The "science question" in feminism has been posed on epistemological, ontological, and ethical registers, drawing in part from foundational critiques of how sex, gender, and nature are associated in modern knowledge.[32] The philosopher and historian Donna Haraway, for example, activated an ethics of feminist thought in a technoscientifically saturated world. Her empirical studies and manifestos steadfastly refused purisms and totalities, uncovering in the sciences hybrids and cyborg amalgams of animals, humans, and machines, opening up reflection on the wide variety of companion species and relationships between humans and nonhumans.[33] As Haraway writes: "In some critical sense that is crudely hinted at by the clumsy category of the social or of agency, the world encountered in knowledge projects is an active entity."[34] Haraway put in place critical elements of a specifically feminist response to this problematic and to this active world, including the "activation" of the previously passive categories of objects of knowledge. This activation had the effect of permanently problematizing binary distinctions like sex and gender without ignoring their strategic utility.[35]

Although recognizing adjacency and partial connection to these two sources for an anthropology of science, this reader delineates a distinctive genealogy of scientific problems and questions. Our aim is not to refute or encompass other efforts in the field of science studies but rather to equip and prepare future anthropologists of the sciences for a field characterized by multiple and competing truth discourses. The work of Marilyn Strathern has consistently rendered visible what experimental forms of anthropology can yield, as well as the danger of the assimilation of heterogeneous discourses into an ecumenical dialogue in anthropology and "in the world." As she wrote in her pathbreaking work *The Gender of the Gift: Problems with Women and Problems with Society in Melanesia*:

> The significance of feminism is the relative autonomy of its premises as far as anthropology is concerned: each provides a critical distance on the other. Ideally, one would exploit the extent to which each talks past the other.[36]

Marilyn Strathern's work likewise exemplifies the challenge faced by anthropologists to wrestle with a simultaneous acknowledgment of their participation in a common world of problems, of the need to take differing intellectual genealogies of thought (and their stakes) seriously, and of the need to recognize their partial connections as well as misunderstandings. Rather than attempting to develop a theory of science that would make everything add up, we present a pathway toward a contemporary anthropology of science.

After attention to mentalities, understood as a heterogeneity of forms and expressions in daily life, and "crises" of representation, either as crises of textual authority or political crises of anthropological practice, we ask: what can we make of a plurality of modes and objects for anthropological inquiry? What can we make of such plurality while retaining a prudent nominalization of our descriptive, analytic, comparative categories "West" and "non-West," "self" and "other," Culture, Society, Humanity, Science, Reason, and Modernity? What is *anthrōpos* amid this plurality?

The task and the challenge, as we understand it, is how to exit the nineteenth-century project of discovering the truth of Man in the empirical details of things without simply creating a situation where all we can do is to track down and transcribe the "actual interconnections among things" in the world, as Max Weber put it, and therewith fail to grasp the "conceptual interconnections among problems."[37] It is the latter task, Weber insisted, which ultimately opens up new insights and significant new points of view. The quest to discover Man has failed in its attempts to bring universal schemata of thought into a stable relation with the variable grammars and experiences of practice, cognitive anthropologies' attention to socioculturally particular schemas notwithstanding.[38] The proliferation of singular ethnographies, however seemingly preferable as an alternative, has frequently failed to deliver on the ever receding promise of establishing meaningful conceptual interconnections. The anthropological orientation of this reader, as we have imagined it, is pragmatic and case based. Pragmatic in that rather than pursuing Man through the empirical, we seek to develop conceptual equipment that will allow us to specify and make sense of problems—specifically anthropological problems of modern science. And case based in that, while cultivating a feel for the partial singularity of things, we also seek concepts that will allow us to put these somewhat singular things into relationship with other things happening in the world. We seek to find conceptual interconnections that will allow us to understand *hetero-logoi* as cases of broader and perhaps more enduring problems.

Such an approach, in our view, is not only exemplified by many of the essays included in this reader; it is exemplified by the reader itself. In selecting and arranging these essays, we have tried to present them as cases of the problem of how to investigate modern science, a problem entailing questions of knowledge and reason, power and authority, ethics and a way of life. In a certain sense, it is the last of these that comes first. The question is not only: What are practices of science today? The question is: How do we become capable of naming anthropological problems of the sciences today? Vocationally, what work on ourselves as anthropologists might we need to do in order to be capable of carrying out this activity?

Our response to the multiple and proliferating renderings of science, as much as to the multiple renderings of *anthrōpos*, is thus neither to seek a unifying theory nor to embrace particularity for its own sake. Rather, we delineate a genealogical pathway in order to provide some of the equipment that will enable readers, we hope, to constitute themselves as subjects of a contemporary anthropology of science.

A Pathway Through Modernities and Sciences

GENEALOGICAL TRAJECTORIES: SCIENCES AND ETHICS

We will now explain why we have selected the particular texts that appear in this reader. We will also explain how we have ordered them and how they form a "genealogical pathway" across a body of philosophical and social scientific works that, in different ways, help us rethink the institutional, cultural, and practical transformations that have defined the sciences and that partially constitute what Michel Foucault called the "modern moment" in the history of truth.

To put this succinctly, we have selected texts that help us engage the sciences as matters of truth, power, and ethics. In our selection and ordering, we have been guided by Foucault's historical inquiries into the ethics and politics of truth and their transformation in modernity. Of particular importance are a series of lecture courses Foucault gave at the Collège de France from 1982 to 1984. In these courses, Foucault traced the early Greek and Roman history of what he called "practices of the self"—practices of self-formation, consisting of philosophical reflection and spiritual exercises, which were carried out in community with the help of friends and mentors. Reflection on these practices of the self, and on the question of formation, proves salutatory when thinking about how the sciences have formed part of modernity and how modern sciences form contemporary life.

Foucault began his historical reflections with the works of Socrates, carried his examination forward to the study of Stoics such as Seneca and Marcus Aurelius, and traced the question of truth and subjectivity into the lives and works of the early Christian monastics. Foucault gave dedicated attention to the manner in which the relation of truth and subjectivity has been deliberately constituted and reconstituted at several key conjunctures in late antiquity and early Christianity in response to changing demands and shifting contexts. One of his primary goals was to clarify what he took to be a major transformation in the relation between truth seeking and ethical practice in the late Middle Ages in Europe. Specifically, he set out to show how prior to what he called "the modern age in the history of truth" philosophers and theologians who wanted to gain access to the truth understood that they would first need to form themselves in and through specific relationships, experiences, and exercises. By contrast, Foucault argued, "the modern age of the history of truth begins when knowledge itself and knowledge alone gives access to the truth" without any prior or ongoing work of ethical formation.[39]

The importance of Foucault's historical insight is that it encourages us to reflect on the fact that for much of the history of the West, truth practices have been deeply connected to matters of power, care, and self-formation—connections that have coalesced, shifted, dissolved, and reformed at various historical junctures and under different configurations. The authors and works presented in this reader provide conceptual tools and critical diagnostics for helping us make sense of those connections as they have been given form in sciences and modernities.

In concordance with research projects initiated by the French scholars Pierre Hadot and Jean-Pierre Vernant on the significance of spiritual exercises for ancient Greek philosophers, Foucault argues that the practice of ancient philosophy was defined by a crucial distinction that has been long since covered over. This distinction, broadly understood, concerns the differences and connections between a classical philosophical mandate to "know oneself" (*gnōthi seauton*) and a second, ethical mandate, regarding the "care for oneself" (*epimeleia heautou*). What's crucial, and what Foucault shows us, is that with the modern moment in the history of truth the relation of priority between these two demands gets inverted. Whereas for ancient thinkers the question of truth (knowing oneself) mattered precisely to the extent that it allowed the thinking subject to live a flourishing life, after the modern moment, the question of knowing oneself, per se, becomes the end and goal of much of philosophy, with the result that it is often assumed that truth is just a matter of knowledge and ethics is just a matter of applying that knowledge effectively.

Foucault helps us make sense of these shifting relations between truth, practice, and subjectivity by offering a conceptual distinction between the term *philosophy* and the term *spirituality*.[40] Philosophy, in Foucault's generative distinction, is not simply "what is true and what is false, but what determines that there is and can be truth and falsehood and whether or not we can separate the true and the false."[41] Philosophy, in other words, is the task of separating the true from the false, a task predicated on the expectation that truth and falsehood can, in fact, be distinguished. By contrast, Foucault defines spirituality as "the set of these researches, practices, and experiences, which may be purifications etc., which are not for knowledge but for the subject, for the subject's very being, the price to be paid for access to the truth."[42] For the sake of pedagogical convenience, Foucault points to a line of thinkers running from Thomas Aquinas to Descartes and Kant as a rough periodization of the shift to a "modern moment" of truth and subjectivity. He is clear, however, that the shift is neither an epochal nor tragic one in the history of ideas. Rather, with the development of what is conventionally referred to as the rise of the "modern world," institutions and practices that provide the bases for the pursuit of truth came to be organized in the forms we now recognize as "scientific research," forms that not only generate a series of new philosophical questions and problems but that tend to cover over the ways in which research is deeply bound up in a form of life.[43]

This reader is organized as a three-part pathway, marking a movement from the troubles of modern research toward contemporary forms of knowledge and living. The texts are assembled to provide equipment for a contemporary anthropology of science, an anthropology that takes the breakdown of knowledge and care as a problem not only for the sciences that are its objects but also for its own knowledge practices.

The first section, "Problems," contains four essays that, in different ways, diagnose breakdowns and pathologies in modern relations of truth, science, and life. These diagnoses focus on the "uses" and venues of reason, the relation of capacities and power relations, and the worth of science in its relation to other dimensions of modern life. Each of the essays is attentive to shifts in the institutional form, practice, and setting of experimental research. They take account of the changing purposes and valuations of scientific practice in relation to the rise of the modern research university, a venue that began to consolidate its form and logic in the early nineteenth century, ultimately transforming the sciences from an avocation of European gentlemen to the institutionalizations and rationalizations of political and industrial enterprise.[44]

The first section of this reader begins with a well-known text written by the German philosopher Immanuel Kant (1724–1804), "An Answer to the Question: What Is Enlightenment?" Kant's text was published in the *Berlinische Monatsschrift* as a response to the question posed a month earlier in the same periodical by the theologian Johann Friedrich Zöllner. Part of the significance of Kant's essay, as the intellectual historian James Schmidt points out, was that it made the question of enlightenment a question of public concern. The question, and its implications for intellectual and political elites, had previously been intensely debated in the Mittwochsgesellschaft (Secret Berlin Wednesday Society), one of the many secret societies in which intellectuals gathered to debate political problems and to discuss reform.[45] The same month as Zöllner's article first posed the question "what is enlightenment?," Johann Karl Möhsen, the personal physician of Friedrich II, the king of Prussia, read a paper to the society (of which Zöllner was a member) in which he noted: members of the Mittwochsgesellschaft could carry out their responsibilities as "well-intentioned patriots" only because "the seal of secrecy" protected them from both the fear of offending patrons and the "thirst for honor or praise."[46]

Both Möhsen's private lecture in the Mittwochsgesellschaft as well as Zöllner's and Kant's public responses underscore the significance of that fact that in the nineteenth century the practice of thinking entailed the creation and expansion of venues for exchange (secret societies and public journals represent just two examples). New questions and new forms of thinking were bound up in new forms of sociality and exchange. It is little wonder that Kant's own response to the question centered on the fraught question of "public" and "private" uses of reason, as well as the ramifications of such uses for intellectual maturity, individual autonomy, and political life in common. For Kant, the private use of reason refers to the use of "reason in a civic post" such as a bureaucrat, soldier, taxpaying citizen, or pastor. In such posts, Kant wrote, one "must obey": the use of reason is highly circumscribed. The soldier following an officer's orders, or the pastor speaking to the congregation, should reason in accord with fixed and preestablished authority relations, regulations, and doctrines. In the public use of reason, however—which Kant describes as making use of one's reason "as a scholar"—*enlightenment* consists precisely in refusing to naively accept parameters and doctrinal boundaries to one's own reasoning, including to the form of its publication.

"Dare to know"—*Sapere Aude!*—Kant implores his reader. The original reference is from Horace's (65 BCE–8 BCE) letter to a soldier, Lollius, to whom he enjoins "*sapere aude, incipe*"—Dare to know, get on with it! If one were to "begin," one would need to think through what form getting on with it would need to take and to decide toward what and with whom one would be getting on with. The task of enlightenment for Kant, in other words, does not presume that a "public" exists as a stable domain or predefined form of activity. Rather, the public use of reason indicates "a certain way of putting to work and using our own faculties."[47] As Foucault describes it, the public use of reason for Kant is:

> the use we make of our understanding and our faculties inasmuch as we
> place ourselves in a universal element in which we can figure as a universal
> subject. . . . We constitute ourselves as a universal subject when as rational
> beings we address all other rational beings. It is simply here, precisely and par
> excellence in that activity of the writer addressing the reader, that we encoun-
> ter a dimension of the public which is at the same time the dimension of the
> universal.[48]

The public use of reason, in short, is a practice, one that entails work on oneself, the creation of venues, and the imagination of a different possible future, in thought and practice, with others.

The first section of our reader ends with Foucault's reflections on the tasks and challenges raised by Kant's essay. Borrowing Kant's title for his own essay, Foucault inflects the question of the public use of reasoning in these terms:

> The question . . . is that of knowing *how* the use of reason can take the public form that it requires, *how* the audacity to know can be exercised in broad daylight, while individuals are obeying as scrupulously as possible.[49]

The importance of these texts for our anthropology is not the reflection on political philosophy per se but rather the importance of the relation of thinking to the identification of breakdowns in thinking and experience in the present and the possibility of inventing a range of practices in response to these discordances. Taking these together, Foucault describes the challenge of thinking under the sign of modernity as one of establishing a "critical anthropology of ourselves." This critical anthropology is marked by a restive attitude toward the present: the challenge of discerning how it is that we as anthropologists can situate ourselves with respect to practices of science, reason, and modern forms of power and ethics. The stakes of inquiry in the present and of the question of our relation as "knowers" to a present that we both seek to understand and participate in are situated in a "paradox of the relation of capacity and power," a relationship that Foucault suggests is far more troubled and subtle "than the eighteenth century may have believed."[50] It is far more troubled and subtle because increased capacities made possible by technologies of different forms of production, reproduction, and self-formation have had as their ends the intensification of certain forms of power relation, powers of regulation, or of normalization, which have frequently been at odds with the freedom that people have sought to exercise over the forms of their lives. The difficulty and subtlety, as Foucault would have it, is then precisely to study those "practical systems" who partake of a polarity between forms of rationality and the manner of possible action. These practical systems, whether we are talking of reproductive technologies, postgenomic biology, or categorizations of mental illness, for example, pose problems of subjectivity, power, and ethics, for participants and observers. As Foucault indicates, inquiry into such problems should take account of the specificity as well as the generality that links the multiplicity and intensity of such problems. It is precisely such generality and specificity of anthropological problems of science, reason, and modernity that we wish to indicate through our pathway.

MODERNITY, KNOWLEDGE, AND CARE: DEWEY AND WEBER

The historian of science Hans-Jörg Rheinberger has recently characterized a turning point in the late nineteenth-century and early twentieth-century relations of historical and positivist sciences as a double crisis of both positivism and of historicism—a "crisis of reality" and an interpretive crisis, respectively.[51] Rheinberger identifies two pivotal episodes generative of the double crisis: "the revolutionary developments in physics, and . . . the problem of the unity of the sciences."[52] This double crisis implicated truth, governance, and ethics alike.

The problem of truth at this juncture could be posed in the following terms: What if the work of knowing never ends? What if the "standards and forms" (to use a phrase from John Dewey) of scientific inquiry change over the course of inquiry? Rheinberger quotes the protozoologist Max Hartmann to the same effect: that science requires the discovery—or invention—of new conceptual systems as demanded by experience.[53]

The crisis of reality, of the unending work of the production of knowledge, had its correlate in an interpretive crisis, perhaps most starkly and somberly characterized by Max Weber shortly before the close of the Great War:

> Scientific works certainly can last as "gratifications" because of their artistic quality, or they may remain important as a means of training. Yet they will be surpassed scientifically—let that be repeated—for it is our common fate and, more, our common goal. We cannot work without hoping that others will advance further than we have. In principle, this progress goes on ad infinitum. And with this we come to inquire into the meaning of science. For, after all, it is not self-evident that something subordinate to such a law is sensible and meaningful in itself. Why does one engage in doing something that in reality never comes, and never can come, to an end?[54]

What, if anything, is the relation between the pursuit of truth claims and the significance of those claims and that pursuit? Contemporaries Weber

(1864–1920) and John Dewey (1859–1952) constitute this shared problem in different terms. A comparison of how these inquirers characterize the problem of science will give us a sense of how inquiry into the significance of science might open up a range of possible responses.

DEWEY: STANDARDS AND FORMS OF INQUIRY

Dewey experienced firsthand the changing conditions of research in the United States at the turn of the twentieth century. He was educated in the 1870s at the University of Vermont, which had recently (1865) been transformed from a private university into a state land-grant institution through the Morrill Act of 1862. The aim of the act was

> without excluding other scientific and classical studies and including military tactic, to teach such branches of learning as are related to agriculture and the mechanic arts, in such manner as the legislatures of the States may respectively prescribe, in order to promote the liberal and practical education of the industrial classes in the several pursuits and professions in life.[55]

In the post–Civil War reconstruction era, utility and public service were considered to be the truly "American" additions to the nineteenth-century German educational ideal of *Bildung*: subject formation. Up until the late 1860s "science" had been seen as a threat to religiously oriented educators in the United States.[56] Accommodations had to be made such that the pursuit of scientific knowledge could be rendered culturally and politically defensible: that is to say, so that these new institutions could gain material support.

As the historians Richard Hofstadter and Walter Metzger show, these accommodations could be characterized as constituting a situation that Weber rigorously described as "value pluralism."[57] An older institutional and organizational unity rooted in religion, exercises of recitation, and classical learning gave way to the proliferation and differentiation of spheres of life with independent norms and forms and, within educational settings, a proliferation of subjects, each considered as equal. Toleration would reign in the American democratic university as long as everyone left everyone else alone. Toleration for diverse subjects of study was frequently justified on the basis of an orientation to the "real." The historian Laurence Veysey has meticulously documented the rise from the 1880s onward of references to "real life" correlated to conceptions of efficiency and efficacy as justifications for the pursuit of diverse forms of knowledge.[58]

Reform of the university in the service of the real had as its aims the cultivation of civic virtue, preparation for work, and the rational solution of public problems. This pluralism was constitutive of a certain conception of the progressive development and worth of research; J. B. Johnson, dean of engineering at the University of Wisconsin, wrote in 1899,

> creature comforts, ante-date culture and sweetness and light are not to be found in squalor or poverty. Scientific agriculture, mining, manufacturing and commerce will, in the future, form the material foundations of all high and noble living.[59]

There was a danger, however, that pluralism married to a materialist conception of progress would descend into value relativism. The danger of relativism was safeguarded by a permanent demand for justification in economic terms. Dewey described the situation in 1902 plainly:

> Institutions of higher learning are ranked by their obvious material prosperity, until the atmosphere of money getting and money spending hides from view the interests for the *sake* of which money alone have a place.[60]

This relation is crucial to Dewey's understanding of education as an ethical and political phenomenon. It is one in which the question of the *sake* for which a practice is done must always be held in view. As Veysey eloquently states the problem of pluralism and judgment:

> A policy of adjustment to "real life" permitted no independent definition of excellence. Indeed it failed even to provide a standard for judging competing definitions of "real life."[61]

For Dewey, the standards through which judgments could be made could only emerge through the process of what he called *inquiry*. Inquiry arises in problematic situations. The "only way out" from a problematic situation, that is, a situation that requires thinking, Dewey observes,

> is through careful inspection of the situation, involving resolution into elements, and a going out beyond what is found upon such inspection to be given, to something else to get a leverage for understanding it. That is we have (a) to

locate the difficulty, and (b) to devise a method for coping with it. Any such way of looking at thinking demands moreover that the difficulty be located in the situation in question.[62]

Dewey was clear on one particularly indeterminate situation, which he named in the second (1948) introduction to his *Reconstruction in Philosophy* (1920):

> the entrance into the conduct of the everyday affairs of life of processes, materials and interests whose origin lies in the work done by physical inquirers in the relatively aloof and remote technical workshops known as laboratories.[63]

The indeterminacy, for Dewey, comes from the incapacity to inquire into and reflect on the effects of the increase in technical means. The consequence of this incapacity to develop an adequate practice of inquiry is that specialists have developed and scoped out domains in distinct jurisdictions, chiefly the material and the ethical. The problematic character of those situations in which this division is instantiated lies in its failure to subject to inquiry our common institutions. Dewey calls for inquiry into institutions funded by public money or that affect diverse publics—institutions that involve ever more elaborate technical control of matter. He calls, furthermore, for inquiry into the habits and morals underlying these ever more elaborate techniques. If one were to do this inquiry, he explains, one would see both logically and morally that the invention of new means does more than alter the ease of achieving the ends we pursue.

Dewey elsewhere stipulates something very important about his understanding of inquiry, through which we can see a resonance with an approach to problems of science as problems including truth, governance, and ethics. For him, inquiry:

> undoubtedly gives an importance to ideas (theories, hypotheses). . . . But it is not a position that can be put in opposition to assertions about matters of particular fact, since, in terms of my view, it states the conditions under which we reach warranted assertability about particular matters of fact. There is nothing peculiarly "pragmatic" about this part of my view, which holds that the presence of an idea—defined as a possible significance of an existent something—is required for any assertion entitled to rank as knowledge or as true; the insistence, however, that the "presence" be by way of an existential operation demarcates it from most other such theories.[64]

Dewey's existential operation was to reconstruct problematic situations, the standard through which he would judge his inquiry. Dewey writes of reconstruction, that it is

> nothing less than the work . . . of forming, of producing (in the literal sense of that word) the intellectual instrumentalities which will progressively direct inquiry into the deeply and inclusively human—that is to say moral—facts of the present scene and situation.[65]

What is pertinent in Dewey's formulation is that science and ethics are interfaced and assembled in accordance with the demands of "progressively directed inquiry." Such a demand is directed toward the possibility of the invention and implementation of intellectual means that facilitate thinking in response to a breakdown in thought and practice.

WEBER: TRUTH AND CONDUCT

Max Weber was one of the most trenchant diagnosticians of modern problems at the turn of the twentieth century. These modern problems, as the historian Fritz Ringer underscores, were identified within a particular mood:

> The social and cultural strains [German industrialization] engendered were unusually severe, and above all, the German academics reacted to the dislocation [of industrialization] with such desperate intensity that the specter of a "soulless" modern age came to haunt everything they said and wrote, no matter what the subject.[66]

Weber diagnosed this specter of the cultural, economic, ethical, and political effects of the "demagification" (*die Entzauberung*) of value-oriented rationalities.[67] Not a reconstructionist with the ambition and hope that science, ethics, and inquiry could be brought into a mutually enriching relation, Weber formulated a tragic diagnosis of modernity's fate. Along with expanding our capacities for knowledge and technical control, modern science had become silent with regards to meaning and morality. Science, Weber claimed, can clarify situations but cannot give us answers to the questions that count most: What should I do? How should I live?[68]

Although Weber saw human beings under conditions of modernity as tragic figures whose knowledge claims could not lead to ethical assurance, this tragedy did not lead him toward a nihilistic or romantic rejection of science. Implicitly evoking the Puritan taskmasters described in his *Protestant Ethic and the Spirit of Capitalism*, he called on scientists to "set to work" and meet the "demands of the day."[69] Precisely the impossibility of a final "answer" motivates the further pursuit of scientific truth, much like the Calvinists who doubled their this-worldly efforts after acknowledging God's absolute power and His absolute deafness to human persuasion. The only possible way in which science could be reunited with ethics, in Weber's diagnosis, was to inhabit science itself as an ethical orientation, to imbue worth and significance into the very practice of scientific work. Thus although science became a profession and the academy ever more professionalized, the ghost of vocation, and the even ghostlier phantom of duty in a calling, nevertheless still haunts us.[70]

Crucially, the significance and worth of pursuing the practice of science cannot be proven in scientific terms; the only question is whether the scientist overextends herself, seeking to answer questions she has no right to answer, or instead puts her life into her work. For Weber (even if he put it ironically), to abandon science altogether and return to the "old churches" did not deserve rebuke; it was those self-styled "academic prophets" who evade "intellectual integrity," using the lecture room as if it were a church pulpit, claiming to prove the scientific necessity of certain value orientations or political allegiances, who deserved the strongest denunciation. The value of science, for Weber as for Dewey, rests in an existential stance that cannot be justified scientifically. If Weber, like other Germans of his time, saw the modern age as "soulless," he also saw a worthy course of life in imbuing the work of science with whatever soul remains inside us.[71]

As a practitioner of what he called the cultural sciences, *Kulturwissenschaften*, whose object is "cultural significance," Weber found himself facing this crisis of modernity in an acute form in the practice of his own work. How could an "objective" and valid science of cultural significance be possible? Weber founds the possibility of objectivity in social science on a primary determination of value necessarily involved in the selection of "objects":

There is no absolutely "objective" scientific analysis of culture—or put perhaps more narrowly but certainly not essentially differently for our purposes—of "social phenomena" independent of special and "one-sided" viewpoints according to which—expressly or tacitly, consciously or unconsciously—they are selected, analyzed and organized for expository purposes.[72]

The "one-sided" view of cultural events is necessary because cultural events are in reality continuous and infinite, which finite human minds cannot grasp in their totality:

> We wish to understand on the one hand the relationships and the cultural significance of individual events in their contemporary manifestations and on the other the causes of their being historically *so* and not *otherwise*. Now, as soon as we attempt to reflect about the way in which life confronts us in immediate concrete situations, it presents an infinite multiplicity of successively and co-existently emerging and disappearing events, both "within" and "outside" ourselves. . . . All the analysis of infinite reality which the finite human mind can conduct rests on the tacit assumption that only a finite portion of this reality constitutes the object of scientific investigation, and that only it is "important" in the sense of being "worthy of being known."[73]

Weber asks: "But what are the criteria by which this segment is selected?" He argues that the attempt to reduce culture to general concepts or laws is "meaningless." Such a reduction is meaningless not because cultural events are lawless anarchy but above all because knowledge of cultural events must depend on the *significance* that concrete constellations of reality have for us. This significance depends not on the "value" of one culture or another but on the fact that we are

> cultured beings, endowed with the capacity and the will to take a deliberate attitude towards the world and to lend it *significance*. Whatever this significance may be, it will lead us to judge certain phenomena of human existence in its light and to respond to them as being (positively or negatively) meaningful.[74]

This meaning cannot be revealed to us in any law; it must come from the value-ideas in the light of which we view "culture" in each individual case. Science cannot lead to meaning or value judgments, but such judgments *necessarily* precede scientific investigation if this science is to lead to any knowledge about cultural phenomena. However, once the object of scientific inquiry is determined through such value judgments, the practice of science and its findings are not subjective "in the sense that they are *valid* for one person and not for others."

In other words, the choice of object of investigation and the extent or depth to which the investigation attempts to penetrate into the infinite causal web, are determined by the evaluative ideas which dominate the investigator and his age. In the *method* of investigation, the guiding "point of view" is of great importance for the *construction* of the conceptual scheme which will be used in the investigation. In the mode of their *use*, however, the investigator is obviously bound by the norms of our thought just as much here as elsewhere. For scientific truth is precisely what is *valid* for all who *seek* the truth.[75]

It follows that scientific truth is itself a value; though valid for all those who *seek* scientific truth, such *seeking* is a historically and culturally specific value all the same. Still, Weber does not approach relativism here for two reasons: first, he argues one must clarify and stand by one's own values rather than tending toward the relativization of all values; second, science involves a *formal* rationality whose *value* may not hold for everyone but whose validity *must* hold.

It has been and remains true that a systematically correct scientific proof in the social sciences, if it is to achieve its purpose, must be acknowledged as correct by a Chinese—or—more precisely stated—it must constantly *strive* to attain this goal, which perhaps may not be completely attainable due to faulty data.[76]

Despite the fact that "a Chinese"—that is, a presumed holder of different values—may find no worth in such a demonstration, it counts as a scientific demonstration only if a Chinese cannot but acknowledge its validity.

This relation between value and science can also be turned back onto the subject. Modernity's loss is certainly real—there is no return to the "age of full and beautiful humanity," the "all-sided" overcoming of the separation of reason and passion sought by Goethe's Faust, which Weber finds as distant from us moderns as are "the clear-eyed Greeks."[77] Instead, one-sidedness—or vocational specialization—must be *embraced* as a modern asceticism, "if it attempts to be a way of life at all, and not simply the absence of any."[78] Embracing not Nietzsche's tragic joy but the sober responsibilities of work and duty in a calling, modernity's fate is infused with vocational meaning. And in doing so, the world too takes on new forms, viewed from ever progressing outlooks and differing standpoints of observation.

TRUTH, SCIENCE, AND THE HUMAN GOOD: AFTER WEBER

"A crisis of our sciences as such: can we seriously speak of it?"[79] The German phenomenologist Edmund Husserl (1859–1938) posed this question at the beginning of his last work, *The Crisis of the European Sciences and Transcendental Philosophy* (1936). The book was the final form given to reflection started in a series of lectures, in Vienna in May 1935 and Prague in November 1935, on what he termed "the European crisis" of man, of science, and of philosophy. The crisis in and of Europe was, for Husserl—after the rise to power of Hitler in Germany—a spiritual, political, scientific, philosophical, and historical crisis.

By 1935, seven years had passed since Husserl's retirement from the Chair of Philosophy at the University of Freiburg (im Breisgau). His retirement had provided Husserl with the opportunity to secure the position for his student and assistant Martin Heidegger (1889–1976). April 1933 marked the beginning of a period in which Husserl would think through the obligation and task of phenomenology: the study of the appearance of things in consciousness and the structures of various kinds of experience, such as desire, memory, bodily experience, and the experience of language. Under the Nazi racial laws of 1933 Husserl was excluded from university life—he had held emeritus status after 1928 and had been continuing his work, engaged in giving lectures. In 1935 he was invited to travel to Vienna and lectured on "Philosophy and the Crisis of European Man." In Prague the "Crisis of European Man" matured into *The Crisis of European Sciences and Transcendental Phenomenology*.

For Husserl science, *Wissenschaft*, is the production of true statements, which are rationally organized. The problem, diagnosed by Weber, was the creation of a division between specialized domains of positive knowledge and "everyday life." Husserl's phenomenology was one response aiming to overcome this divide. Such a phenomenology, as articulated in the *Crisis*, did not have as its concern the scientific character of the sciences but what sciences can mean for human existence (*menschliches Dasein*). The problem of the meaning of philosophy was urgent because of the historical conjuncture of Europe in the 1930s. Husserl wrote, under the heading "The history of modern philosophy as a struggle for the meaning of man,"

If we consider the effect of the development of philosophical ideas on (non-philosophizing) mankind as a whole, we must conclude the following: Only an understanding from within the movement of modern philosophy from Descartes to the present, which is coherent despite all its contradictions, makes possible an understanding of the present itself. The true struggles of our time, the only ones which are significant, are struggles between humanity which has already collapsed and humanity which still has roots but is struggling to keep them or find new ones. The genuine spiritual struggles of European humanity as such take the form of struggles between the skeptical philosophies—or nonphilosophies that retain the word but not the task—and the actual and still vital philosophies. But the vitality of the latter consists in the fact that they are struggling for their own true and genuine meaning and thus for the meaning of a genuine humanity.[80]

Husserl's 1935 rethinking of the task of phenomenology, by attending to the *historical* stakes of the breakdown between truth and life, marks a transition to the works by Hans Blumenberg and Georges Canguilhem included in the second section of this reader, "Historical Problematizations." Husserl attempted to reconstruct the relation and significance of reason (*logos*) for living. In line with our orienting problem formulation from Dewey and Weber, Husserl characterized the relation between science and life in his day as a problematically declining ratio of the capacity to know and the significance of that knowledge:

The exclusiveness with which the total world-view of modern man, in the second half of the nineteenth century let itself be determined by the positive sciences and be blinded by the "prosperity" they produced, meant an indifferent turning away from the questions which are decisive for a genuine humanity. Merely fact-minded sciences make merely fact-minded people.[81]

As though answering Weber's question as to whether science can still be of use for life for the one who poses questions correctly, the *Crisis* proposed the possibility of a future philosophy and of a different relation of science to living.

Husserl argues in the *Crisis* that a philosophy adequate to the task of the present must engage in historical inquiry. By this he did not mean an empirical history of science as it developed historically but rather a reflection back from the present to "what was originally and always sought in philosophy"—the grasping of the phenomenal lifeworld (*Lebenswelt*) through concepts.

We require historical analysis of concepts, according to Husserl, because reflection on the inheritance of these concepts has been neglected, leading to their "sedimentation." The sedimentation of concepts through forgetfulness has the effect that the original meaning of such concepts is concealed from us. The meaning of concepts, such as those of geometry, were "originally" rooted in the lifeworld—the practices and ends of experiencing subjects. For Husserl, crisis comes with the divergence of the world disclosed by science from the phenomenal lifeworld. The crisis of the 1930s was the historical development of a scientific lifeworld that made it seem as though the phenomenal lifeworld was irrelevant to knowledge of phenomena.[82]

The proposal of the *Crisis* is a "transcendental" history of scientific concepts. The aim is not an empirical history of science but rather a reconstruction of the paradigmatic events constitutive of the conceptual apparatus of the sciences. The core turning points in Husserl's judgment come with Galileo and the mathematicization of the physics of an infinite universe. Husserl was concerned that the terms of a mathematical characterization of natural phenomena resulted in an incapacity to pose questions of meaning.

If Husserl's problem was to make the transcendental standpoint historical, his former student Martin Heidegger's problem was the epochal withholding of Being under conditions of modernity.

In his 1938 essay "The Age of the World Picture," he wrote:

> The fundamental event of the modern age is the conquest of the world as picture. The word "picture" [*Bild*] now means the structured image [*Gebild*] that is the creature of man's producing which represents and sets before. In such producing, man contends for the position in which he can be that particular being who gives the measure and draws up the guidelines for everything that is. Because this position secures, organizes, and articulates itself as a world view, the modern relationship to that which is, is one that becomes, in its decisive unfolding, a confrontation of world views; and indeed not of random world views, but only of those that have already taken up the fundamental position of man that is most extreme, and have done so with the utmost resoluteness. For the sake of this struggle of world views and in keeping with its meaning, man brings into play his unlimited power for the calculating, planning, and molding of all things. Science as research is an absolutely necessary form of this establishing of self in the world; it is one of the pathways upon which the modern age rages toward fulfillment of its essence, with a velocity unknown to the participants. With this struggle of world views the modem age first enters into

the part of its history that is the most decisive and probably the most capable of enduring.[83]

Heidegger's project for fundamental ontology, a phenomenology of Being as such, and not of particular beings, was thus a search for a mode of countermodern thinking under conditions of a diagnosis and prognosis of modernity.[84]

REPROBLEMATIZATIONS

Included in the second section of the reader, "Historical Problematizations," is the work of the historian of ideas Hans Blumenberg, who offers the most compelling reproblematization of the conceptual development of Husserl's and Heidegger's formulations of problems of science under conditions of modernity, the legitimacy of scientific curiosity about the world, and the consequences of such a position and *attitude* for how one could study scientific practice.

Blumenberg (1920–1996) developed a unique approach to the history and historicity of truth. This approach, which comes broadly under the banner of histories of metaphor and concepts, investigates the history and intertwining of nonconceptuality (*Unbegrifflichkeit*) and concept formation. For the early Blumenberg, especially in his thesis work—*The Ontological Distance: An Inquiry Into the Crisis of Husserl's Phenomenology*—the translator and commentator Robert Savage explains, "history unfolds in the interspace between the two limit terms or poles of the 'ontological distance' which he labels 'oppositionality' (*Gegenständigkeit*) and 'extrapositionality' (*Inständigkeit*)."[85] The former is the pole of a subject grounded in method, exemplified by Husserl (and Descartes), through which both knowledge of the world and action in the world are justified. The latter, exemplified by Heidegger's analysis of Dasein, is the pole "in which human existence spurns the assurances offered to it by reason, religion, and tradition to face up to its own contingency."[86] Savage highlights the fact that "Blumenberg marks his own position by vindicating the 'fallen' or 'inessential' realm of history against those who would transcend it in either direction."[87]

History for Blumenberg is a movement space (*Bewegungsraum*) that insofar as it enters a modern moment becomes a milieu of human self-affirmation. In Blumenberg's first major work, *The Legitimacy of the Modern Age*, from which our reading is selected, this milieu and this attitude toward human capacities for thought and action, as well as thought about thought and action, is identified as an original and legitimate characteristic of modernity. As Blumenberg puts it, "it is curiosity that draws one's attention to curiosity."[88] Whereas Husserl understands the "theoretical attitude" as a transhistorical, European constant, Blumenberg locates the originality of curiosity in its modern manifestation precisely in a renunciation of "theoretical" continuity; a refusal to answer prior questions with later means (and vice versa). This renunciation is also a renunciation of justification—the requirement that curiosity be legitimated in terms other than self-affirmation of the curious attitude.

KNOWLEDGE, LIFE, AND CARE

For the French historical epistemologists, from Gaston Bachelard to Georges Canguilhem, the history of science is an inquiry into the historicity of truth and reason. The history of science is therefore not akin to a history of ideas, in which ideas of each historical period are taken up relative to that period's cultural norms. Rather, the fundamental problem is how something can be both true and historical.[89] On the one hand, the standards of truth and rationality can be found only in the historical achievements of the sciences.[90] On the other hand, science is characterized by a fundamental break with everyday life, such that the history of science develops differently from the history of other domains of practice.[91]

Whereas Bachelard focused his historical inquiry into the physical and chemical sciences, Canguilhem adopted and adapted Bachelard's approach for the history of the life and medical sciences. Indeed, for Canguilhem it is life—and the knowledge of life—that forms the central problem for a history of truth. The objectivity of medical and biological truth are grounded in the "living" of organisms. Any attempt to explain a living organism as a mechanical product of the material conditions of its environment, for example, founders on the organism's active self-constitution of its environment and its relationships. As Canguilhem writes in "The Living and Its Milieu," drawing on the biology of Jakob von Uexküll and Kurt Goldstein,

The relation between the living and the milieu establishes itself as a debate (*Auseinandersetzung*), to which the living brings its own proper norms of appreciating situations, both dominating the milieu and accommodating itself to it. This relation does not essentially consist (as one might think) in a struggle, in an opposition. That applies to the pathological state. A life that affirms itself against the milieu is a life already threatened. . . . A healthy life, life confident in its existence, in its values, is a life of flexion, suppleness, almost softness.[92]

Furthermore, science itself is an activity undertaken by a living being. The problem of a knowledge of life is not only a problem posed by living beings to objective knowledge but also the problem posed by the very fact of a living being that knows. As Canguilhem puts it, "as a living being, man does not escape the general law of living beings."[93] Even the most objective forms of knowledge are, at their origins, the work of living beings and therefore maintain a "permanent and obligatory relationship with perception," and we would add, with judgment.[94] Yet, as Foucault has noted, Canguilhem's "living" (*vivant*) is not the phenomenologist's "lived experience" (*vécu*) but rather remains focused on the *problematic* relation between reason and existence, knowledge and life.

In an interview in 1981, shortly after beginning his transition from work on pastoral power and modern political reason toward a problematization of knowledge and care, Foucault described a lineage extending from Bachelard, Jean Cavaillès, and Canguilhem into his own projects on truth. In order to connect his own work to those who came before him, Foucault asked:

through the analysis of collective social historical experiences, tied to specific historical contexts, how can we do the history of a knowledge [*l'histoire d'un savoir*], the history of our sciences [*l'histoire de nos connaissances*], and how could new objects show up [*arriver*] in the domain of knowledge, how could they present themselves as objects to be known?[95]

There is significant distance separating Foucault's work on "dubious" knowledge domains (such as psychiatry or criminology) from Bachelard's and Canguilhem's concern for the specific conceptual order and truth practices made possible in the natural sciences. Nevertheless, the core point of intersection we wish to render visible and activate is the investigation (and not denunciation or reification) of the processes by which truth is elaborated. Such processes have been investigated through the norms of rationality at play, norms that are historically and geographically specific and that involve relations of power. If in the above citation Foucault articulates a common attention to the historical emergence of "objects" in the domain of knowledge, his later work turned attention to the question of the historicity of the subject, extending Canguilhem's concern with the relation between knowledge and the living. Foucault's later work marks out the pathway from "Historical Problematizations" toward the third section of this reader, "Ethics: Subjectivity and Truth."

In Foucault's later work, particularly in his studies of ancient sexuality in the years before his untimely death, Foucault intensified his inquiry into ethics, which he understood as a study of the modes by which the subject actively and freely constitutes a relation to itself (*rapport à soi*)—what Foucault called modes of subjectivation. As he writes in the second volume of *The History of Sexuality*:

> I will start from the common notion of "use of the pleasures"—*chrēsis aphro-disiōn*—and attempt to determine the modes of subjectivation to which it referred: the ethical substance, the types of subjection, the forms of elaboration of the self, and the moral teleology.[96]

The term *chrēsis* is an important one: it denotes both "use" of something, the capacity and power of using as well the intimacy and participation with the object of engagement. The term highlights the ethical work involved in what might otherwise be understood as "mere" instrumentality. In working sessions with Rabinow and the philosopher Hubert Dreyfus in Berkeley, Foucault gave an orientation to his problematization of ethics by explicating the four major aspects in which the relation of a subject to itself is constituted, four aspects that James Faubion has clarified (as well as extended) as parameters of the ethical domain:[97] "substance"—which is to say the object of reflection and practice; the "mode of subjectivation"—which is to say, how subjects are ethically qualified, recognizing ethical obligations and identifying ethical capacities and incapacities; the *telos* or ends of that practice—such ends could be internal or external to such practice; as well as *askēsis*—the ethical exercise of work on the specified substance, toward an end, and to become that subject of an ethically qualified kind.

Foucault's attention to practices of becoming an ethical subject included asking how subjects become capable of speaking the truth. Discussing the ancient Greek problematization of *parrhēsia*—"frank speech"—Foucault suggests a turn towards an "alethurgic" rather than epistemological notion of truth. Instead of investigating the conditions under which a discourse can be recognized as true, an "alethurgic" study analyzes "the form in which, in his act of telling the truth, the individual constitutes himself and is constituted by others as a subject of a discourse of truth."[98] For example, Foucault contrasts the parrhēsiast with the professional rhetor, the sage, and the teacher. Although the contents and even epistemological rules of the truths spoken by these four figures may be identical, they differ in terms of "the conditions and forms of the type of act by which the subject manifests himself when speaking the truth, by which I mean, thinks of himself and is recognized by others as speaking the truth."[99] Truth under the conditions of the modern sciences, for Foucault, is problematic not because of the form of knowledge but because of the form taken by the subject that speaks the truth. As Foucault puts it,

> we can say that the modern age of the relations between the subject and truth begin when it is postulated that, such as he is, the subject is capable of the truth, but that, such as it is, the truth cannot save the subject.[100]

The problem of the subject and truth that Foucault introduced and with respect to which we situate our anthropological problem of science is thus not epistemological or ontological per se but rather ethical. And as we turn to our concluding section, we reiterate that this ethical problem and task of becoming a subject of truth under the conditions of modernity bears as much on the anthropologist of the sciences as on her objects of study.

Conclusion

By assembling these texts into a pathway, we hope to provide readers with something more than a collection of disparate (or even antagonistic) possible understandings of science and modernity. Rather, we hope that the genealogical form of the pathway allows readers to draw conceptual interconnections among the problems posed in each text. Perhaps more importantly, we hope that the pathway will help readers establish their own position of critical adjacency to—rather than inside of or opposed to—modern problematizations of science. This adjacency is a position that draws from but is not determined by the historicity of modern problems of science. It is a position that facilitates a contemporary anthropology of science. We draw the concept of the "contemporary" from the recent work of Paul Rabinow.

The contemporary, Rabinow writes, is "a moving ratio of modernity, moving through the recent past and the near future in a (nonlinear) space that gauges modernity as an ethos already becoming historical."[101] Drawing from this insight, we assemble the texts included here as a contemporary movement through the modern problematizations of science and life. Drawing conceptual equipment from the recent past, the pathway provides routes toward future anthropologies of the sciences. The modern problematizations of science, those troubled figures illuminated by Dewey and Weber, Canguilhem and Foucault, are themselves already becoming historical. But it is only by taking up, observing, and reflecting on one's own relationship to these problems of modern science that future anthropologists of science can articulate lines of inquiry neither bound within nor forgetful of the modern crises of truth and life. In that spirit we conclude the reader with a selection from Rabinow's 2003 conceptual anthropological meditation *Anthropos Today: Reflections on Modern Equipment*, a reading that is very much contemporary insofar as it scopes out such an adjacent position and such an ethos.

This reader strives to be as much invitational as instructive; in making this pedagogical legacy available to a wider body of students and scholars in anthropology and science studies, it invites its readers to situate themselves and their concerns along this same genealogical line. Along the way, it provides readers with now-classic responses to persistent questions—How can and should the human sciences engage with other practices and practitioners of scientific knowledge? What forms of moral reflection and political analysis are appropriate to such inquiry? What standards of judgment might fruitfully be put to use? What does such inquiry require in terms of self-clarification and the cultivation of new scholarly capacities? Equally important, it offers an occasion to make these questions one's own.

We put these readings together and pursued this double aim—to equip and to invite—in the spirit of philosophical friendship. Over the past number of years the three of us, in different ways and to different ends, have undertaken the sometimes difficult and often joy-filled labor of placing our own work and pedagogical formation in this same legacy. And, like the reader, we too were invited to strive to make anthropological sense of modern science and practices of reason and to make such anthropological striving available to others. To put it in terms borrowed from Rabinow, we have been the beneficiaries of "modern equipment" for observing, thinking about, and participating in the modern sciences; we would like, as they say, to pay it forward.

Problems

An Answer to the Question: "What Is Enlightenment?"

Immanuel Kant

Immanuel Kant (1724–1804) is one of the central figures in modern philosophy. His critical enterprise aimed to identify and test the nature and limits of reason and experience. In addition to his major philosophical works on reason, history, ethics, and politics, he also wrote numerous timely and occasional essays. "An Answer to the Question: 'What Is Enlightenment?'" was the second of fifteen such essays to be published in the *Berlinische Monatsschrift*, a monthly periodical and public forum whose founding editors were members of the secretive Berlin Wednesday Society (*Berliner Mittwochsgesellschaft*), a small group of political, religious, philosophical, and economic thinkers committed to questions of enlightenment and reform. Kant's essay was written in response to the question posed a month earlier in the same periodical by the theologian Johann Friedrich Zöllner, amid a swirling and sometimes combative debate about intellectual freedom and the nature and limits of authority. The essay proposes a diagnosis of the distinctive capacities and responsibilities that characterize the goal of modern enlightenment. It has become a defining statement of the ethics and aspirations of modern thought.

Enlightenment is man's emergence from his self-incurred immaturity. Immaturity is the inability to use one's own understanding without the guidance of another. This immaturity is *self-incurred* if its cause is not lack of understanding, but lack of resolution and courage to use it without the guidance of another. The motto of enlightenment is therefore: *Sapere aude!*[1] Have courage to use your *own* understanding!

Laziness and cowardice are the reasons why such a large proportion of men, even when nature has long emancipated them from alien guidance (*naturaliter maiorennes*),[2] nevertheless gladly remain immature for life. For the same reasons, it is all too easy for others to set themselves up as their guardians. It is so convenient to be immature! If I have a book to have understanding in place of me, a spiritual adviser to have a conscience for me, a doctor to judge my diet for me, and so on, I need not make any efforts at all. I need not think, so long as I can pay; others will soon enough take the tiresome job over for me. The guardians who have kindly taken upon themselves the work of supervision will soon see to it that by far the largest part of mankind (including the entire fair sex) should consider the step forward to maturity not only as difficult but also as highly dangerous. Having first infatuated their domesticated animals, and carefully prevented the docile creatures from daring to take a single step without the leading-strings to which they are tied, they next show them the danger which threatens them if they try to walk unaided. Now this danger is not in fact so very great, for they would certainly learn to walk eventually after a few falls. But an example of this kind is intimidating, and usually frightens them off from further attempts.

Thus it is difficult for each separate individual to work his way out of the immaturity which has become almost second nature to him. He has even grown fond of it and is really incapable for the time being of using his own understanding, because he was never allowed to make the attempt. Dogmas and formulas, those mechanical instruments for rational use (or rather misuse) of his natural endowments, are the ball and chain of his permanent immaturity. And if anyone did throw them off, he would still be uncertain about jumping over even the narrowest of trenches, for he would be unaccustomed to free movement of this kind. Thus only a few, by cultivating their own minds, have succeeded in freeing themselves from immaturity and in continuing boldly on their way.

There is more chance of an entire public enlightening itself. This is indeed almost inevitable, if only the public concerned is left in freedom. For there will always be a few who think for themselves, even among those appointed as guardians of the common mass. Such guardians, once they have themselves thrown off the yoke of immaturity, will disseminate the spirit of rational respect for personal value and for the duty of all men to think for themselves. The remarkable thing about this is that if the public, which was previously put under this yoke by the guardians, is suitably stirred up by some of the latter who are incapable of enlightenment, it may subsequently compel the guardians themselves to remain under the yoke. For it is very harmful to propagate prejudices, because they finally avenge themselves on the very people who first encouraged them (or whose predecessors did so). Thus a public can only achieve enlightenment slowly. A revolution may well put an end to autocratic despotism and to rapacious or power-seeking oppression, but it will never produce a true reform in ways of thinking. Instead, new prejudices, like the ones they replaced, will serve as a leash to control the great unthinking mass.

For enlightenment of this kind, all that is needed is *freedom*. And the freedom in question is the most innocuous form of all—freedom to make *public use* of one's reason in all matters. But I hear on all sides the cry: *Don't argue!* The officer says: Don't argue, get on parade! The tax-official: Don't argue, pay! The clergyman: Don't argue, believe! (Only one ruler in the world says: *Argue* as much as you like and about whatever you like, *but obey!*).[3] All this means restrictions on freedom everywhere. But which sort of restriction prevents enlightenment, and which, instead of hindering it, can actually promote it? I reply: The *public use* of man's reason must always be free, and it alone can bring about enlightenment among men; the *private use* of reason may quite often be very narrowly restricted, however, without undue hindrance to the progress of enlightenment. But by the public use of one's own reason I mean that use which anyone may make of it *as a man of learning* addressing the entire *reading public*. What I term the private use of reason is that which a person may make of it in a particular *civil* post or office with which he is entrusted.

Now in some affairs which affect the interests of the commonwealth, we require a certain mechanism whereby some members of the commonwealth must behave purely passively, so that they may, by an artificial common agreement, be employed by the government for public ends (or at least deterred from vitiating them). It is, of course, impermissible to argue in such cases; obedience is imperative. But in so far as this or that individual who acts as part of the machine also considers himself as a member of a complete commonwealth or even of cosmopolitan society, and thence as a man of learning who may through his writings address a public in the truest sense of the word, he may indeed argue without harming the affairs in which he is employed for some of the time in a passive capacity. Thus it would be very harmful if an officer receiving an order from his superiors were to quibble openly, while on duty, about the appropriateness or usefulness of the order in question. He must simply obey. But he cannot reasonably be banned from making observations as a man of learning on the errors in the military service, and from submitting these to his public for judgment. The citizen cannot refuse to pay the taxes imposed upon him; presumptuous criticisms of such taxes, where someone is called upon to pay them, may be punished as an outrage which could lead to general insubordination. Nonetheless, the same citizen does not contravene his civil obligations if, as a learned individual, he publicly voices his thoughts on the impropriety or even injustice of such fiscal measures. In the same way, a clergyman is bound to instruct his pupils and his congregation in accordance with the doctrines of the church he serves, for he was employed by it on that condition. But as a scholar, he is completely free as well as obliged to impart to the public all his carefully considered, well-intentioned thoughts on the mistaken aspects of those doctrines, and to offer suggestions for a better arrangement of religious and ecclesiastical affairs. And there is nothing in this which need trouble the conscience. For what he teaches in pursuit of his duties as an active servant of the church is presented by him as something which he is not empowered to teach at his own discretion, but which he is employed to expound in a prescribed manner and in someone else's name. He will say: Our church teaches this or that, and these are the arguments it uses. He then extracts as much practical value as possible for his congregation from precepts to which he would not himself subscribe with full conviction, but which he can nevertheless undertake to expound, since it is not in fact wholly impossible that they may contain truth. At all events, nothing opposed to the essence of religion is present in such doctrines. For if the clergyman thought he could find anything of this sort in them, he would not be able to carry out his official duties in good conscience, and would have to resign. Thus the use which someone employed as a teacher makes of his reason in the presence of his congregation is purely *private*, since a congregation, however large it is, is never any more than a domestic gathering. In view of this, he is not and cannot be free as a priest, since he is acting on a commission imposed from outside. Conversely, as a scholar addressing the real public (i.e., the world at large) through his writings, the clergyman making *public use* of his reason enjoys unlimited freedom to use his own reason and to speak in his own person. For to maintain that the guardians of the people in spiritual matters should themselves be immature, is an absurdity which amounts to making absurdities permanent.

But should not a society of clergymen, for example an ecclesiastical synod or a venerable presbytery (as the Dutch call it), be entitled to commit itself by oath to a certain unalterable set of doctrines, in order to secure for all time a constant guardianship over each of its members, and through them over the people? I reply that this is quite impossible. A contract of this kind, concluded with a view to preventing all further enlightenment of mankind forever, is absolutely null and void, even if it is ratified by the supreme power, by Imperial Diets and the most solemn peace treaties. One age cannot enter into an alliance on oath to put the next age in a position where it would be impossible for it to extend and correct its knowledge, particularly on such important matters, or to make any progress whatsoever in enlightenment. This would be a crime against human nature, whose original destiny lies precisely in such progress. Later generations are thus perfectly entitled to dismiss these agreements as unauthorized and criminal. To test whether any particular measure can be agreed upon as a law for a people, we need only ask whether a people could well impose such a law upon itself. This might well be possible for a specified short period as a means of introducing a certain order, pending, as it were, a better solution. This would also mean that each citizen, particularly the clergyman, would be given a free hand as a scholar to comment publicly, i.e., in his writings, on the inadequacies of current institutions. Meanwhile, the newly established order would continue to exist, until public insight into the nature of such matters had progressed and proved itself to the point where, by general consent (if not unanimously), a proposal could be submitted to the crown. This would seek to protect the congregations who had, for instance, agreed to alter their religious establishment in accordance with their own notions of what higher insight is, but it would not try to obstruct those who wanted to let things remain as before. But it is absolutely impermissible to agree, even for a single lifetime, to a permanent religious constitution which no-one might publicly question. For this would virtually nullify a phase in man's upward progress, thus making it fruitless and even detrimental to subsequent generations. A man may for his own person, and even then only for a limited period, postpone enlightening himself in matters he ought to know about. But to renounce such enlightenment completely, whether for his own person or even more so for later generations, means violating and trampling underfoot the sacred rights of mankind. But something which a people may not even impose upon itself can still less be imposed upon it by a monarch; for his legislative authority depends precisely upon his uniting the collective will of the people in his own. So long as he sees to it that all true or imagined improvements are compatible with the civil order, he can otherwise leave his subjects to do whatever they find necessary for their salvation, which is none of his business. But it is his business to stop anyone forcibly hindering others from working as best they can to define and promote their salvation. It indeed detracts from his majesty if he interferes in these affairs by subjecting the writings in which his subjects attempt to clarify their religious ideas to governmental supervision. This applies if he does so acting upon his own exalted opinions—in which case he exposes himself to the reproach: *Caesar non est supra Grammaticos*,[4] but much more so if he demeans his high authority so far as to support the spiritual despotism of a few tyrants within his state against the rest of his subjects.

If it is now asked whether we at present live in an *enlightened* age, the answer is: No, but we do live in an age of *enlightenment*. As things are at present, we still have a long way to go before men as a whole can be in a position (or can ever be put into a position) of using their own understanding confidently and well in religious matters, without outside guidance. But we do have distinct indications that the way is now being cleared for them to work freely in this direction, and that the obstacles to universal enlightenment, to man's emergence from his self-incurred immaturity, are gradually becoming fewer. In this respect our age is the age of enlightenment, the century of *Frederick*.[5]

A prince who does not regard it as beneath him to say that he considers it his duty, in religious matters, not to prescribe anything to his people, but to allow them complete freedom, a prince who thus even declines to accept the presumptuous title of *tolerant*, is himself enlightened. He deserves to be praised by a grateful present and posterity as the man who first liberated mankind from immaturity (as far as government is concerned), and who left all men free to use their own reason in all matters of conscience.

Under his rule, ecclesiastical dignitaries, notwithstanding their official duties, may in their capacity as scholars freely and publicly submit to the judgment of the world their verdicts and opinions, even if these deviate here and there from orthodox doctrine. This applies even more to all others who are not restricted by any official duties. This spirit of freedom is also spreading abroad, even where it has to struggle with outward obstacles imposed by governments which misunderstand their own function. For such governments can now witness a shining example of how freedom may exist without in the least jeopardizing public concord and the unity of the commonwealth. Men will of their own accord gradually work their way out of barbarism so long as artificial measures are not deliberately adopted to keep them in it.

I have portrayed *matters of religion* as the focal point of enlightenment, i.e., of man's emergence from his self-incurred immaturity. This is firstly because our rulers have no interest in assuming the role of guardians over their subjects so far as the arts and sciences are concerned, and secondly, because religious immaturity is the most pernicious and dishonorable variety of all. But the attitude of mind of a head of state who favors freedom in the arts and sciences extends even further, for he realizes that there is no danger even to his *legislation* if he allows his subjects to make *public* use of their own reason and to put before the public their thoughts on better ways of drawing up laws, even if this entails forthright criticism of the current legislation. We have before us a brilliant example of this kind, in which no monarch has yet surpassed the one to whom we now pay tribute.

But only a ruler who is himself enlightened and has no fear of phantoms, yet who likewise has at hand a well-disciplined and numerous army to guarantee public security, may say what no republic would dare to say: *Argue as much as you like and about whatever you like, but obey!* This reveals to us a strange and unexpected pattern in human affairs (such as we shall always find if we consider them in the widest sense, in which nearly everything is paradoxical). A high degree of civil freedom seems advantageous to a people's intellectual freedom, yet it also sets up insuperable barriers to it. Conversely, a lesser degree of civil freedom gives *intellectual* freedom enough room to expand to its fullest extent. Thus once the germ on which nature has lavished most care—man's inclination and vocation to *think freely*—has developed within this hard shell, it gradually reacts upon the mentality of the people, who thus gradually become increasingly able to *act freely*. Eventually, it even influences the principles of governments, which find that they can themselves profit by treating man, who is *more than a machine*,[6] in a manner appropriate to his dignity.*

* I read today on the 30th September in Büsching's[7] *Wochentliche Nachrichten* of 13th September a notice concerning this month's *Berlinische Monatschrift*. The notice mentions Mendelssohn's[8] answer to the same question as that which I have answered. I have not yet seen this journal, otherwise I should have held back the above reflections. I let them stand only as a means of finding out by comparison how far the thoughts of two individuals may coincide by chance.

Science as a Vocation

Max Weber

Max Weber (1864–1920) is one of the major figures in the development
of the modern social sciences. He is considered to be one of the foremost
diagnosticians of the character and history of modernity. His lecture
"Science as a Vocation" describes the ethical challenge of living a life
devoted to science within modern bureaucratic institutions and under
conditions of modern capitalism. The essay was delivered to students
of science and engineering late in World War I. At the time, Germany
was broken by war but not yet defeated. In the midst of horrific political,
social, and economic destruction, and aware of the pressing question of
Germany's role in instigating that destruction, Weber offered a trenchant
and surprising meditation on the worth of a life devoted to science—a life
whose significance is not primarily found in the industrial and economic
outcomes of research but rather from a sense of inward calling to a life of
inquiry and honest self-clarification. Weber's statement remains one of the
most compelling critiques of, and invitations to, the life of science.

It is your wish that I should talk about "science as a vocation."[1] Now, we
political economists possess a certain pedantic streak that I should like to
retain. It is expressed in the fact that we always start from external circumstances. In this instance this means starting with the question: What form
does science take as a profession in the material sense of the word? In practical terms this amounts nowadays to the question: What is the situation of
a graduate student who is intent on an academic career in the university?
In order to understand the particular nature of circumstances in Germany

it will be helpful to proceed comparatively and to see how matters stand abroad, above all in the United States, which in this respect presents the sharpest possible contrast with us.

As everyone knows, here in Germany the career of a young man who chooses science as a profession normally begins as a "lecturer" [*Privatdozent*]. After consulting with and gaining the approval of a representative of the relevant discipline, he qualifies[2] as a university lecturer on the basis of a book and an examination—something of a formality for the most part—in the presence of the faculty as a whole. He then gives lectures on topics of his own choosing within the limits of the *venia legendi*, his license to teach. For this he receives no salary, and he is rewarded only with the lecture fees paid by his students.[3] In America an academic career normally begins quite differently, namely, with an appointment as an "assistant." This is similar to what happens in Germany in the large institutes of the natural sciences and medicine, where the second doctorate, which is the formal qualification of a lecturer, is obtained only by a fraction of the assistants, and then often only late in their careers. The difference means in practice that in Germany an academic career is generally based on plutocratic premises. For it is extremely risky for a young scholar without private means to expose himself to the conditions of an academic career. He must be able to survive at least for a number of years without knowing whether he has any prospects of obtaining a position that will enable him to support himself. The United States, in contrast, has a bureaucratic system. A young man receives a salary from the outset—a modest one, to be sure. His salary barely amounts to the wages of a worker one rung above an unskilled laborer. Even so, having a fixed salary, he begins with an apparently secure position. However, as a rule, he can be dismissed, like our assistants, and frequently he must reckon that the authorities will not hesitate to dismiss him if he fails to meet their expectations. What is expected is that he will achieve "full houses." This cannot happen to a German *Privatdozent*. Once you have him, there is no getting rid of him. It is true that he has no "rights." But he does have the understandable expectation that if he has worked for years on end he has a kind of moral claim to consideration. This includes being considered—and this is frequently important—in the context of the possible appointment of other lecturers. This raises the question of whether on principle every competent scholar should be allowed to qualify, or whether "teaching needs" should be taken into account. Since this effectively gives the existing lecturers a teaching monopoly, a painful dilemma arises that is closely related to the dual aspect of the academic profession, which will be discussed shortly. For the most part, the second option is chosen. But that increases the risk that however conscientious he may be subjectively, the relevant department head will end up giving preference to his own students. Personally, I should make it clear that I have always followed the principle that a scholar whom I have supervised for his Ph.D. should apply to *someone else* to study for the second doctorate and thus legitimate himself elsewhere. But as a consequence one of my best students found himself rejected by another university since no one would *believe* that this was my reason.

There is a further difference between America and Germany. This is that in Germany the lecturer is *less* concerned with lecturing than he might wish. He does indeed have the right to lecture on any topic in his discipline. But to make use of that right is thought to show an unseemly lack of respect toward lecturers with greater seniority, and as a rule the "major" lectures are given by the professor as the departmental representative of the discipline while the lecturer makes do with ancillary lectures. The advantage of this is that he can devote his early years to research, even though he may not do so entirely voluntarily.

In America the system is organized on entirely different principles. In his early years the young lecturer is completely overloaded precisely because he is *paid*. In a department of German studies, for example, the full professor will give a three-hour course of lectures a week on, say, Goethe, and that is all, while the junior university assistant will have twelve hours teaching a week, including the duty of drumming the basics of German grammar into students' heads, and he will be happy if he is assigned the task of lecturing on writers up to the rank of, say, Uhland.[4] For the syllabus is prescribed by the departmental authorities and the assistant is as dependent on them as the institute assistant is in Germany.

Now we can see very clearly that the latest developments across broad sectors of the German university system are moving in the same direction as in America. The major institutes of science and medicine are "state-capitalist" enterprises. They cannot be administered without funding on a huge scale. So we see the situation that exists wherever capitalist operations are to be found, namely, the "separation of the worker from the means of production." The worker, in this instance the assistant, is dependent on the resources that are provided by the state. He is as dependent on the institute director, therefore, as an employee in a factory is dependent on his boss—for the institute director believes in good faith that this institute is *his* institute and that it is his to manage. The assistant's situation, then, is as precarious as that of every "quasi-proletarian" existence and as that of an assistant[5] in an American university.

Our German university life is becoming Americanized in very important respects, as is German life in general. I am convinced that this development will continue to spread to disciplines like my own where the artisan is still the owner of his own resources (which amount essentially to the library), just as the old craftsman of the past owned the tools of his trade. This development is in full swing.

Its technical advantages are beyond doubt, as is the case with all capitalist and bureaucratized activities. But the "spirit" that prevails in them is different from the traditional climate of German universities. Both outwardly and inwardly, a vast gulf separates the head of a large capitalist university enterprise of this kind and the average old-style full professor. This applies also to their inner attitude, though I cannot go into that here. Both in essence and appearance, the old *constitution* of the university has become a fiction. What has remained and has even been radically intensified is a feature peculiar to a university *career*. This is the fact that for a lecturer, let alone an assistant, to succeed in rising to the position of a full professor or even the head of an institute is purely a matter of *luck*. Chance is not the only factor, but its influence is quite exceptional. I know of scarcely any other profession on earth where it plays such a crucial role. I feel at liberty to make this claim since I personally owe it to a number of purely chance factors that I was appointed to a full professorship while still very young[6] in a discipline in which people of my own age had undoubtedly achieved more than I. And it is this experience that encourages me to believe that I have developed a keen eye for the undeserved fate of the many whom chance has treated, and continues to treat, in the opposite way and who have failed, for all their abilities, to obtain a position that should rightfully be theirs through this selection process.

That chance, rather than ability, plays such an important role, is not exclusively or even chiefly the product of the human factors that are just as prevalent in the selection process in universities as in any other. It would be unjust to blame personal shortcomings in either faculties or the Ministries of Education for the fact that so many mediocrities occupy leading positions in our universities. The cause is to be sought instead in the laws governing human cooperation, especially the cooperation of a number of different bodies, in this instance, the proposing faculties and the ministries.[7] By way of comparison we can observe the events that have taken place over many centuries in the course of papal elections: the most important verifiable example of a comparable selection process. It is rare for the cardinal who is said to be the "favorite" to have any prospects of success. As a rule, the second or third candidate on the list is selected. The same may be said of the president of the United States. Only exceptionally does the first-rate, outstanding candidate manage to obtain the "nomination" of the party conventions and subsequently run in the election. Mainly it is the number two or number three man. The Americans have already devised technical sociological expressions for all these categories, and it would be interesting to use these examples to study the laws governing this process of selection through the formation of a collective will. However, we cannot do this today. But these laws also apply to university staff, and what is astonishing is not that mistakes are often made, but that, despite everything, the number of *good* appointments is relatively large. Only where parliaments intervene for *political* reasons, as happens in a number of countries, can we be sure that only safe mediocrities or careerists will have prospects of obtaining appointments. The same thing may be said of countries like Germany, where monarchs interfered for similar reasons and where, at present, revolutionary leaders do likewise.

No university teacher likes dwelling on the discussions that precede the filling of posts, for they are seldom pleasant. And yet I can say that in the numerous cases known to me, the sincere *intention* to reach decisions on purely objective grounds was always present without exception.

For we must make a further attempt at clarification. The fact that chance plays such a major role in deciding academic destinies does not spring from the defects of collective decision making as a part of the selection process. Every young man who feels he has a vocation as a scholar must be aware that the task awaiting him has a dual aspect. He must be properly qualified not only as a scholar, but also as a teacher. And these two things are by no means identical. A man can be both an outstanding scholar and an execrable teacher. I may remind you of the teaching activities of such men as Helm-holtz or Ranke.[8] And these are far from being isolated cases. Now the pres-ent situation is that our German universities, especially the smaller ones, are caught up in a ludicrous popularity contest. The local landlords in our university towns celebrate the arrival of the thousandth student with a party but would like to welcome the two thousandth with a torchlight procession. "Crowd-pleasing" appointments in neighboring disciplines have a consider-able impact on lecture fees, and we should be quite frank about this. And even if we leave that aside, the number of enrolled students is a statistically tangible proof of success, whereas the qualities of a scholar are imponder-able and frequently (and very naturally) a matter of dispute, particularly in the case of bold innovators.

For this reason almost everyone succumbs to the idea that large student numbers are a blessing and a value in their own right. If a lecturer is said to be a bad teacher, this amounts in most cases to an academic death warrant, even if he is the greatest scholar in the world. But the question of whether an academic is a good teacher or a bad one is answered with reference to the frequency with which students honor him with their presence. However, it is also true that the fact that students flock to a teacher is determined largely by purely extraneous factors such as his personality or even his tone of voice—to a degree that might scarcely be thought possible.

After extensive experience and sober reflection on the subject, I have de-veloped a profound distrust of lecture courses that attract large numbers, unavoidable though they may be. Democracy is all very well in its right-ful place. In contrast, academic training of the kind that we are supposed to provide in keeping with the German university tradition is a matter of *aristocratic spirit*, and we must be under no illusions about this. On the other hand, it is quite true that perhaps the most challenging pedagogic task of all is to explain scientific problems in such a way as to make them comprehen-sible to an untrained but receptive mind, and to enable such a person—and this is the only decisive factor for us—to think about them independently. There can be no doubt about this, but it is not student numbers that decide whether this task has been accomplished. And—to return to our theme— the art of teaching is a personal gift and does not necessarily coincide with a scholar's qualities as a researcher. Unlike France, however, we have no body comprising the "Immortals" of learning, while in the German tradition the universities are supposed to do justice to both tasks, research and teaching. But whether the talents needed for this can be united in a single individual is a matter of pure chance.

Thus academic life is an utter gamble. When young students come to me to seek advice about qualifying as a lecturer, the responsibility of giving it is scarcely to be borne. Of course, if the student is a Jew, you can only say: *lasciate ogni speranza.*[9] But others, too, must be asked to examine their conscience: Do you believe that you can bear to see one mediocrity after another being promoted over your head year after year, without your becoming embittered and warped? Needless to say, you always receive the same answer: of course, I live only for my "vocation"—but I, at least, have found only a handful of people who have survived this process without injury to their personality.

So much for the external conditions of a scholarly vocation.

But I believe that you really want to hear about something else, about an *inner* vocation for science. At the present time, that inner vocation, in contrast to the external organization of science as a profession, is determined in the first instance by the fact that science has entered a stage of specialization that has no precedent and that will continue for all time. Not just outwardly, but above all inwardly, the position is that only through rigorous specialization can the individual experience the certain satisfaction that he has achieved something perfect in the realm of learning. With every piece of work that strays into neighboring territory, work of the kind that we occasionally undertake and that sociologists, for example, must necessarily produce, we must resign ourselves to the realization that the best we can hope for is to provide the expert with useful *questions* of the sort that he may not easily discover for himself from his own vantage point inside his discipline. Our own work, however, will inevitably remain highly imperfect. Only rigorous specialization can give the scholar the feeling for what may be the one and only time in his entire life, that here he has achieved something that will *last*.

Nowadays, a really definitive and valuable achievement is always the product of specialization. And anyone who lacks the ability to don blinkers for once and to convince himself that the destiny of his soul depends upon whether he is right to make precisely this conjecture and no other at this point in his manuscript should keep well away from science. He will never be able to submit to what we may call the "experience" of science. In the absence of this strange intoxication that outsiders greet with a pitying smile, without this passion, this conviction that "millennia had to pass before you were born, and millennia more must wait in silence" to see if your conjecture will be confirmed—without this you do *not* possess this vocation for science and should turn your hand to something else. For nothing has any value for a human being as a human being unless he *can* pursue it with *passion*.

Nevertheless, the fact remains that however genuine and profound such a passion may be, it is a far cry from guaranteeing success. Passion is, of course, a precondition of the decisive factor, namely, "inspiration." Among young people nowadays the idea is very widespread that science has become a question of simple calculation, something produced in laboratories or statistical card indexes, just as "in a factory," with nothing but cold reason and not with the entire "soul." Though of course we should note in passing that for the most part there is very little understanding of what actually goes on in a factory or a laboratory. In both places it is necessary for something, and the right thing at that, to *occur* to people if they are to achieve anything worthwhile.

But inspiration cannot be produced to order. And it has nothing in common with cold calculation. Undoubtedly, calculation, too, is an unavoidable prerequisite. For example, no sociologist, even when advanced in years, should think himself too high and mighty to spend months on end doing tens of thousands of quite trivial sums in his head. You cannot shift the burden entirely to mechanical aids with impunity if you want to achieve anything, and what you do achieve is often little enough. But if you do not have a definite idea about the purpose of your calculation, and if during the calculation nothing "occurs" to you about the implications of the individual answers as they arise, then even that "little" will fail to appear. Normally, inspiration flourishes only on a foundation of very hard work. Not always, of course. The inspiration of an amateur can be as productive scientifically as that of an expert, or even more so. We owe many of our very best methods of tackling problems and our best insights to amateurs. The only difference between an amateur and an expert is, as Helmholtz observed about Robert Mayer,[10] that the amateur lacks a tried and tested method of working. He is therefore mainly not in a position to judge or evaluate or pursue the implications of his inspiration. Inspiration does not do away with the need for work.

And for its part, work cannot replace inspiration or force it to appear, any more than passion can. Both work and passion, and especially both *together*, can entice an idea. Ideas come in their own good time, not when we want them. In fact, the best ideas occur to us while smoking a cigar on the sofa, as Ihering[11] says, or during a walk up a gently rising street, as Helmholtz observes of himself with scientific precision, or in some such way. At any rate, ideas come when they are least expected, rather than while you are racking your brains at your desk. But by the same token, they would not have made their appearance if we had not spent many hours pondering at our desks or brooding passionately over the problems facing us.

However that may be, the scholar must resign himself to the element of chance that is involved in every kind of scientific endeavor. It is expressed in the question: Will inspiration come or not? A man may be an outstanding worker and yet never have had a valuable idea of his own. But it is a grave error to imagine that this is true only of science and that in an office, for example, the situation is different from a laboratory. A businessman or a big industrialist without "commercial imagination," that is to say, without inspiration or brilliant ideas will continue his whole life long to be someone who ought rather to be a clerk or a technical official. He will never introduce organizational innovations. It is not at all the case—as academic conceit would have us believe—that inspiration plays a greater role in science than in the solving of the problems of practical life by the modern entrepreneur. And on the other hand, people often fail to recognize that inspiration does not play a smaller part in science than in the realm of art. It is childish to imagine that a mathematician will arrive at any kind of valuable scientific discoveries by sitting at a desk with a ruler or other mechanical tools or calculators. The mathematical imagination of a Weierstrass[12] is, of course, organized very differently both in its meaning and its consequences from that of an artist, and indeed, there is a fundamental difference in quality. But not in terms of the psychological process involved. Both are intoxication (in the sense of Plato's "mania")[13] and "inspiration."

Now, whether someone has scientific inspiration depends on fates that are hidden from us, but also on "talent." It is not least this indisputable truth that has led to a belief that, understandably enough, is particularly popular among young people. Today, that belief has put itself at the service of a number of idols whose shrines are to be found today at every street corner and in every periodical. These idols are "personality" and "experience," and the two are closely connected. The idea is prevalent that experience forms the essence of personality and is an integral part of it. People put themselves through torture in order to "experience" things, for that is an essential part of the proper lifestyle of a "personality," and if they do not succeed they must at the very least try to act as if they possessed this gift of grace. Formerly, this "experience" [*Erlebnis*] was known in German as "sensation" [*Sensation*]. And I believe that the latter term provided a more accurate idea of what "personality" is and means.

Ladies and gentlemen, in the realm of science, the only person to have "personality" is the one who is *wholly devoted to his subject*. And this is true not just of science. We know of no great artist who has ever done anything other than devoted himself to his art and to that alone. Even a personality of Goethe's stature had to pay a price, as far as his art was concerned, for having taken the liberty of trying to turn his "life" into a work of art. And even if you question that this was his aim, you at least have to be Goethe to take that liberty. Moreover, it will surely be admitted that even a man like him, who appears only once in a thousand years, could not emerge from this wholly unscathed. In politics things are no different, but that cannot be discussed here today. Even in the realm of science, however, we may say categorically that if a man appears on the stage as the impresario of the subject to which he devotes himself and if he attempts to legitimate himself by appealing to his "personal experience," this is not enough to turn him into a personality. Nor is it the sign of a personality to go on to ask: How can I show that I am more than just a mere "expert"? How can I manage to prove that I can say something in form or substance, that no one has ever said? This phenomenon has increased massively nowadays and always seems petty. It always diminishes the man who asks such questions instead of allowing his inner dedication to his task and to it alone to raise him to the height and the dignity of the cause he purports to serve. And in this respect, the situation with the artist is no different.

These preconditions of our work are factors that we share with art. But we now find them confronted with a destiny that opens up a vast gulf between science and artistic endeavors. Scientific work is harnessed to the course of *progress*. In the realm of art, however, there is no such thing as progress in that sense. It is untrue that a work of art that is created in an age which has developed new techniques, such as the laws of perspective, is somehow superior in purely artistic terms to a work of art that is innocent of all such techniques and laws. At least, such a work of art is not inferior as long as it does justice to its own form and materials, in other words, if it selects and shapes its object in a way that is appropriate even without those laws and techniques. A work of art that truly achieves "fulfillment" will never be surpassed; it will never grow old. The individual can assess its significance for himself personally in different ways. But no one will ever be able to say that a work that achieves genuine "fulfillment" in an artistic sense has been "superseded" by another work that likewise achieves "fulfillment."

Contrast that with the realm of science, where we all know that what we have achieved will be obsolete in ten, twenty, or fifty years. That is the fate, indeed, that is the very *meaning* of scientific work. It is subject to and dedicated to this meaning in quite a specific sense, in contrast to every other element of culture of which the same might be said in general. Every scientific "fulfillment" gives birth to new "questions" and *cries out* to be surpassed and rendered obsolete. Everyone who wishes to serve science has to resign himself to this. The products of science can undoubtedly remain important for a long time, as "objects of pleasure" because of their artistic qualities, or as a means of training others in scientific work. But we must repeat: to be superseded scientifically is not simply our fate but our goal. We cannot work without living in hope that others will advance beyond us. In principle, this progress is infinite.

This brings us to the *problem of the meaning of science*. For it is far from self-evident that a thing that is subject to such a law can itself be meaningful and rational. What is the point of engaging in something that neither comes, nor can come, to an end in reality? Well, for one thing, we may engage in it for purely practical purposes, or technical purposes in a broader sense: namely, to enable us to orient our practical actions by the expectations provided by our scientific experience. All well and good. However, that has meaning only for the practical man. But what is the inner attitude of the scientist himself to his profession? If indeed he bothers to search for one. He maintains that science must be pursued "for its own sake," and not simply so that others can use it to achieve commercial or technical successes, so that they can feed or clothe themselves, make light for themselves, or govern themselves. What meaningful achievement can he hope for from activities that are always doomed to obsolescence? What can justify his readiness to harness himself to this specialized, never-ending enterprise? That question calls for some general reflections.

Scientific progress is a fraction, and indeed the most important fraction, of the process of intellectualization to which we have been subjected for thousands of years and which normally provokes extremely negative reactions nowadays.

Let us begin by making clear what is meant in practice by this intellectual process of rationalization through science and a science-based technology. Does it mean, for example, that each one of us sitting here in this lecture room has a greater knowledge of the conditions determining our lives than an Indian or a Hottentot? Hardly. Unless we happen to be physicists, those of us who travel by streetcar have not the faintest idea how that streetcar works. Nor have we any need to know it. It is enough for us to know that we can "count on" the behavior of the streetcar. We can base our own behavior on it. But we have no idea how to build a streetcar so that it will move. The savage has an incomparably greater knowledge of his tools. When we spend money, I would wager that even if there are political economists present in the lecture room, almost every one of them would have a different answer ready to the question of how money manages things so that you can sometimes buy a lot for it and sometimes only a little. The savage knows how to obtain his daily food and what institutions enable him to do so.

Thus the growing process of intellectualization and rationalization does *not* imply a growing understanding of the conditions under which we live. It means something quite different. It is the knowledge or the conviction that if *only we wished* to understand them we *could* do so at any time. It means that in principle, then, we are not ruled by mysterious, unpredictable forces, but that, on the contrary, we can in principle *control everything by means of calculation*. That in turn means the disenchantment of the world. Unlike the savage for whom such forces existed, we need no longer have recourse to magic in order to control the spirits or pray to them. Instead, technology and calculation achieve our ends. This is the primary meaning of the process of intellectualization.

Let us consider this process of disenchantment that has been at work in Western culture for thousands of years and, in general, let us consider "progress," to which science belongs both as an integral part and a driving force.

Can we say that it has any meaning over and above its practical and technical implications? This question has been raised on the level of principle in the works of Leo Tolstoy. He arrived at the problem by a curious route. What he brooded about increasingly was whether or not *death* has a meaning. His answer was that it had no meaning for a civilized person. His reasoning for this was that because the individual civilized life was situated within "progress" and infinity, it could not have an intrinsically meaningful end. For the man caught up in the chain of progress always has a further step in front of him; no one about to die can reach the pinnacle, for that lies beyond him in infinity. Abraham or any other peasant in olden times died "old and fulfilled by life"[14] because he was part of an organic life cycle, because in the evening of his days his life had given him whatever it had to offer and because there were no riddles that he still wanted to solve. Hence he could have "enough" of life. A civilized man, however, who is inserted into a never-ending process by which civilization is enriched with ideas, knowledge, and problems may become "tired of life," but not fulfilled by it. For he can seize hold of only the minutest portion of the new ideas that the life of the mind continually produces, and what remains in his grasp is always merely provisional, never definitive. For this reason death is a meaningless event for him. And because death is meaningless, so, too, is civilized life, since its senseless "progressivity" condemns death to meaninglessness. This idea pervades all of Tolstoy's late novels,[15] and it defines the keynote of art.

How should we respond to this? Does "progress" as such possess a recognizable meaning that goes beyond the technical so that devotion to progress can become a meaningful vocation? This question cannot be avoided. But it ceases to be merely a question of a vocation *for* science, in other words, the problem of the meaning of science as a career for the person who chooses it. Instead, it turns into the question of what is the *vocation of science* within the totality of human life? And what is its value?

There is a vast gulf here between past and present. You will recall the marvelous image at the beginning of Book 7 of Plato's *Republic*. He describes there the cavemen in chains with their gaze directed at the wall of rock in front of them. Behind them lies the source of light that they cannot see; they see only the shadows the light casts on the wall, and they strive to discover the relationship between them. Until one of them succeeds in bursting his bonds and he turns around and catches sight of the sun. Blinded, he stumbles around, stammering about what he has seen. The others call him mad. But gradually he learns to look into the light, and his task then is to clamber down to the cavemen and lead them up into the light of day. He is the philosopher, while the sun is the truth of science, which alone does not snatch at illusions and shadows but seeks only true being.

Well, who regards science in this light today? Nowadays, the general feeling, particularly among young people, is the opposite, if anything. The ideas of science appear to be an otherworldly realm of artificial abstractions that strive to capture the blood and sap of real life in their scrawny hands without ever managing to do so. Here in life, however, in what Plato calls the shadow theater on the walls of the cave, we feel the pulse of authentic reality; in science we have derivative, lifeless will-o'-the-wisps and nothing else. How did this turnabout take place? Plato's passionate enthusiasm in the *Republic* is ultimately to be explained by the fact that for the first time the meaning of the *concept* had been consciously discovered, one of the greatest tools of all scientific knowledge. It was Socrates who discovered its implications. He was not alone in this respect. You can find very similar approaches in India to the kind of logic developed by Aristotle. But nowhere was its significance demonstrated with this degree of consciousness. In Greece for the first time there appeared a tool with which you could clamp someone into a logical vise so that he could not escape without admitting either that he knew nothing or that this and nothing else was the truth, the *eternal* truth that would never fade like the actions of the blind men in the cave. That was the tremendous insight of the pupils of Socrates. And it seemed to follow from this that once you had discovered the correct concept for the beautiful, the good, or, let us say, courage, or the soul, or whatever it might be, you would have grasped its true nature. And this appeared to be the key to knowing and to teaching people how to act rightly in life, above all, as citizens. For this was the crucial issue for the Greeks, whose thought was political through and through. And that explains why science was a worthwhile activity.

This discovery by Greek philosophy was now joined during the period of the Renaissance by the second great tool of scientific work. This was rational experiment as a way of controlling experience reliably, without which modern empirical science would be impossible. There had been earlier experiments. For example, physiological experiments had been conducted in India in connection with the ascetic techniques of the Yogi, mathematical experiments for military purposes in ancient Greece, and there had also been experiments in the Middle Ages in such fields as mining. But to have elevated the experiment to the principle of research as such was the achievement of the Renaissance. The pioneers here were the great innovators in the realm of *art*, like Leonardo and his contemporaries. Of particular importance were the musical experimenters of the sixteenth century with their experimental keyboards. Starting from these men, the experiment migrated into science above all through Galileo, and it entered theory with Bacon. After that, it was adopted by the exact sciences in continental universities, beginning with Italy and the Netherlands.

What did science mean to these people on the threshold of modernity? For artistic experimenters like Leonardo and the musical innovators of the sixteenth century, it meant the path to *true* art, and for them this meant the path to true *nature*. Art should be elevated to the rank of a science, and this meant, above all, that the artist should be raised to the rank of a doctor,[16] both socially and in terms of the meaning of his life. That, for instance, was the ambition underlying Leonardo's notebooks. And today? "Science as the path to nature"—that would be blasphemy in the ears of modern youth. No, it is the other way around. Young people today want release from the intellectualism of science in order to return to their own nature and hence to nature as such! And science as the way to art? Criticism is superfluous. But even more was expected of science in the age of the emergence of the exact natural sciences. Remember the statement by Jan Swammerdam: "I bring you the proof of God's providence in the anatomy of a louse."[17] You can see from this how scientific work conceived of its own task under the (indirect) influence of Protestantism and Puritanism. It thought of science as the way to God. That way was no longer to be discovered by the philosophers with their concepts and deductions. The fact that God could no longer be found where the Middle Ages had looked for him was known to the entire theology of Pietism of the day, Spener above all.[18] God is hidden, his ways are not our ways, his thoughts are not our thoughts. In the exact natural sciences, however, where his works could be experienced physically, people cherished the hope that they would be able to find clues to his intentions for the world.

And today? Apart from the overgrown children who can still be found in the natural sciences, who imagines nowadays that a knowledge of astronomy or biology or physics or chemistry could teach us anything about the *mean-*

ing of the world? How might we even begin to track down such a "meaning," if indeed it exists? If anything at all, the natural sciences are more likely to ensure *that* the belief that the world has a "meaning" will wither at the root! And in particular, what about the idea of science as the path "to God"? Science, which is specifically alien to God? And today no one can really doubt in his heart of hearts that science is alien to God—whether or not he admits it to himself. Release from the rationalism and intellectualism of science is the fundamental premise of life in communion with the divine.

This, or something very like it, is one of the basic slogans that you hear from our young people who are religiously minded or in search of religious experience. And they are in search not just of religious experience, but of experience as such. The only surprising thing is the path they take. This is that the only realm that intellectualism had failed to touch until now, namely, the realm of the irrational, is what is now made conscious and subjected to intellectual scrutiny. For that is what the modern intellectualist romanticism of the irrational amounts to in practice. This method of liberating us from the intellect brings about the exact opposite of what is envisaged by those who adopt it. Thus a naive optimism had led people to glorify science, or rather the techniques of mastering the problems of life based on science, as the road to *happiness*. But after Nietzsche's annihilating criticism of those "last men" "who have discovered happiness,"[19] I can probably ignore this completely. After all, who believes it—apart from some overgrown children in their professorial chairs or editorial offices?

Let us return to our theme. Given these internal assumptions, what is the meaning of science as a vocation now that all these earlier illusions—"the path to true existence," "the path to true art," "the path to true nature," "the path to the true God," "the path to true happiness"—have been shattered? The simplest reply was given by Tolstoy with his statement, "Science is meaningless because it has no answer to the only questions that matter to us: 'What should we do? How shall we live?'"[20] The fact that science cannot give us this answer is absolutely indisputable. The question is only in what sense does it give "no" answer, and whether or not it might after all prove useful for somebody who is able to ask the right question. People are wont to speak nowadays of a science "without presuppositions." Does such a thing exist? It depends on what is meant by it. Every piece of scientific work presupposes the validity of the rules of logic and method. These are the fundamental ways by which we orient ourselves in the world. Now, there is little to object to in these presuppositions, at least for our particular question. But science further assumes that the knowledge produced by any particular piece of scientific research should be *important*, in the sense that it should be "worth knowing." And it is obvious that this is the source of all our difficulties. For this presupposition cannot be proved by scientific methods. It can only be interpreted with reference to its ultimate meaning, which we must accept or reject in accordance with our own ultimate attitude toward life.

Furthermore, the relationship of scientific research to these presuppositions varies according to their structure. Sciences such as physics, chemistry, and astronomy presuppose as self-evident that it is worth knowing the ultimate laws governing cosmic processes insofar as they can be scientifically construed. Not simply because this can lead to technical advances, but, if science is supposed to be a "vocation," "for their own sake." This presupposition cannot itself be proved. Even less can we show that the world that these laws describe deserves to exist, that it has a "meaning" and that it is meaningful to live in it. These sciences do not ask such questions.

Or, take the example of a practical art like modern medicine, which is so highly developed in scientific terms. The general "presupposition" of medical practice is, to put it trivially, that its task is to preserve life as such and to reduce suffering as far as possible. And that is problematic. The doctor uses all his scientific skill to keep alive a dying man even if he begs to be released from this life, and even if his relatives wish for, and must wish for, the patient's death, whether they admit it or not, because his life is worthless, because they do not begrudge him his release from suffering and because they find that the expense of maintaining his worthless existence has become unbearable—he may well be a wretched madman. But the presuppositions of medicine and the penal code prevent the doctor from desisting from his efforts. Whether this life is valuable and when, medical science does not inquire. All natural scientists provide us with answers to the question: what should we do if we wish to *make use of technology* to control life? But whether we wish, or ought, to control it through technology, and whether it ultimately makes any sense to do so, is something that we prefer to leave open or else to take as a given.

Or consider a discipline like aesthetics and art history. The fact that works of art exist is a given. Aesthetics seeks to explain the conditions in which they arise. But it does not inquire whether the realm of art may not in fact be a realm of diabolic magnificence, a kingdom of this world and hence intrinsically inimical to God and, given its profoundly aristocratic spirit, hostile to human fellowship. It does not ask whether works of art *should* exist.

Or, again, take jurisprudence. This examines the body of legal thought that has been built partly on logic and partly on practices established by convention. It determines which elements are valid; in other words, it determines *when* specific rules of law and specific modes of interpretation are to be recognized as authoritative. It does not explain *whether* such a thing as law should exist and *whether* these particular rules should be adopted. Jurisprudence can only tell us that if we wish for success, then according to the norms of our legal system the best way to achieve it is to apply this particular rule of law.

Or consider the different branches of cultural history. They teach us how to understand the political, artistic, literary, and social products of culture by examining the conditions that gave rise to them. But they provide no answer to questions about whether these cultural products *deserved* or deserve to exist. Nor do they answer the other question of whether it is worth taking the trouble to get to know them. They assume that we have an interest in using this procedure to establish our membership in the community of "civilized human beings." But whether this is the case in reality is not something they can demonstrate "scientifically," and the fact that they presuppose it does not at all imply that it is self-evident. Because that is far from being the case.

Let us now turn to the disciplines familiar to me, that is to say, sociology, history, economics, and political science, and the branches of philosophy that are concerned with interpreting them. It is often said, and I subscribe to this view, that politics has no place in the lecture room. It has no place there as far as students are concerned. I would, for example, disapprove just as much if pacifist students were to make their appearance in the lecture room of my former colleague Dietrich Schäfer[21] in Berlin, surround the lectern, and make the sort of commotion said to have been created by anti-pacifist students during a lecture given by Professor Foerster,[22] a man whose opinions are in many respects as remote from my own as it is possible to be. But it is likewise true that politics has no place in the lecture room as far as the lecturer is concerned. Least of all if his subject is the academic study of politics. For opinions on issues of practical politics and the academic analysis of political institutions and party policies are two very different things. If you speak about democracy at a public meeting there is no need to make a secret of your personal point of view. On the contrary, you have to take one side or the other explicitly; that is your damned duty. The words you use are not the tools of academic analysis, but a way of winning others over to your political point of view. They are not plowshares to loosen the solid soil of contemplative thought, but swords to be used against your opponents: weapons, in short.

In a lecture room it would be an outrage to make use of language in this way. When we speak of democracy in the course of a lecture, our task is to examine its various forms, to analyze them in order to see how they work, and to establish the consequences of this or that version for people's lives. We should then compare them with nondemocratic political systems. Our aim must be to enable the listener to discover the vantage point from which *he* can judge the matter in the light of *his* own ultimate ideals. But the genuine teacher will take good care not to use his position at the lectern to promote any particular point of view, whether explicitly or by suggestion. For this latter tactic is, of course, the most treacherous approach when it is done in the guise of "allowing the facts to speak for themselves."

Now, why should we not do this? I may start by saying that many highly esteemed colleagues of mine are of the opinion that it is not possible to act in accordance with this self-denying ordinance, and if it were possible it would simply be a cranky notion that were best avoided. Now we cannot provide a university teacher with scientific proof of where his duty lies. All we can demand of him is the intellectual rectitude to realize that we are dealing with two entirely *heterogeneous* problems. On the one hand, we have the establishing of factual knowledge, the determining of mathematical or logical relations or the internal structure of cultural values. On the other, we have answers to questions about the *value* of culture and its individual products, and in addition, questions about how we should *act* in the civilized community and in political organizations. If he then asks why he cannot deal with both sets of problems in the lecture room, we should answer that the prophet and the demagogue have no place at the lectern. We must say to both the prophet and the demagogue: "go out into the street and speak to the public."[23] In other words, speak where what you say can be criticized. In the lecture room, where you sit opposite your listeners, it is for them to keep silent and for the teacher to speak.

I think it irresponsible for a lecturer to exploit a situation in which the students have to attend the class of a teacher for the sake of their future careers but where there is no one present who can respond to him critically. It is irresponsible for such a teacher to fail to provide his listeners, as is his duty, with his knowledge and academic experience, while imposing on them his personal political opinions. No doubt, an individual lecturer will not always be able to suppress his subjective sympathies. He will then have to face the sharpest criticism in the forum of his own conscience. And it proves nothing, for other, purely factual errors are possible and yet they do not amount to a refutation of the idea that his duty is to seek the truth. Furthermore, I reject the idea in the interests of pure science. I am willing to demonstrate from the writings of our historians that whenever an academic introduces his own value judgment, a complete understanding of the facts comes to an end. But this goes beyond the limits of the theme of my lecture this evening and would call for lengthy explanations.

I ask only this: suppose that we give a class on the forms of church and the state or on the history of religion to a group that includes a practicing Catholic on the one side, and a Freemason on the other. And if we do, how shall we attempt to persuade them to agree to the same *evaluation*? It is quite impossible. And yet the academic teacher must wish and must demand of himself that he should be of use to both of them through his knowledge and his grasp of method. Now you will have every right to say that even in a factual account of the events leading to the emergence of Christianity, a devout Catholic will never be willing to accept the view of a teacher who does not share his dogmatic preconceptions. That is undoubtedly true! But the difference consists in this. Science, which is without "preconceptions" in the sense that it rejects any religious allegiance, likewise has no knowledge of "miracles" and "revelation." If it did, it would be untrue to its own "preconceptions." The religious believer has knowledge of both. And a science without "preconceptions" expects of the believer no less, but also *no more* than the recognition that *if* the course of events can be explained without recourse to supernatural interventions that must be excluded from an empirical account of the causal factors involved, then it will have to be explained in the way that science attempts to do so. And that is something the believer can do without compromising his faith.

But we may go on to ask whether the achievements of science have no meaning for anyone who is indifferent to facts as such and is interested only in the practical point of view. Perhaps they do after all. To make an initial point: the first task of a competent teacher is to teach his students to acknowledge *inconvenient* facts. By these I mean facts that are *inconvenient* for their own personal political views. Such extremely inconvenient facts exist for every political position, including my own. I believe that when the university teacher makes his listeners accustom themselves to such facts, his achievement is more than merely intellectual. I would be immodest enough to describe it as an "ethical achievement," though this may be too emotive a term for something that is so self-evident.

Up to now, I have spoken only of *practical* reasons for not imposing one's personal opinions on others. But we must go further. There are much deeper reasons that persuade us to rule out the "scientific" advocacy of practical points of view—except, that is, for the discussion of what means to choose in order to achieve an end that has been definitely *agreed*. Such advocacy is senseless in principle because the different value systems of the world are caught up in an insoluble struggle with one another. The elder Mill, whose philosophy I do not otherwise admire, was right on this one point when he said that if you take pure experience as your starting point, you will end up in polytheism. This is to put it superficially and it sounds paradoxical, but it contains some truth. If we know anything, we have rediscovered that something can be sacred not just although it is not beautiful, but *because* and *insofar* as it is not beautiful. Evidence of this can be found in the book of Isaiah, chapter 53, and in Psalm 21.[24] And we know that something can be beautiful not just although it is not good but even in the very aspect that lacks goodness. We have known this ever since Nietzsche, and the same message could be gleaned earlier in the *Fleurs du mal*—as Baudelaire entitled his volume of poems. And it is a truism that something can be true although and because it is neither beautiful nor sacred, nor good. But these are merely the most basic instances of this conflict between the gods of the different systems and values.

I do not know how you would go about deciding "scientifically" between the *value* of French and German culture. Here, too, conflict rages between different gods and it will go on for all time. It is as it was in antiquity before the world had been divested of the magic of its gods and demons, only in a different sense. Just as the Greek would bring a sacrifice at one time to Aphrodite and at another to Apollo, and above all, to the gods of his own city, people do likewise today. Only now the gods have been deprived of the magical and mythical, but inwardly true qualities that gave them such vivid immediacy. These gods and their struggles are ruled over by fate, and certainly not by "science." We cannot go beyond understanding *what* the divine means for this or that system or within this or that system. And this spells the end of any discussion by professors in lecture rooms, although, of course, the great problems of *life* implicit here is far from being exhausted.

But forces other than the holders of university chairs are at work here. What man will take it upon himself to provide a "scientific refutation" of the morality of the Sermon on the Mount, and in particular its dictum "Resist not him that is evil" or the metaphor of turning the other cheek?[25] And yet it is clear that, regarded from a worldly point of view, what is being preached here is an ethics of ignoble conduct. We must choose between the religious dignity that this ethics confers and the human code of honor [*Manneswürde*] that preaches something altogether different, namely, "Resist evil, otherwise you will bear some of the responsibility for its victory." According to his point of view, each individual will think of one as the devil and the other as God, and he has to decide which one is the devil and which the God *for him*. And the same thing holds good for all aspects of life. The awe-inspiring rationalism of a systematic ethical conduct of life that flows from every religious prophecy dethroned this polytheism in favor of the "One thing that is needful."[26] Then, when confronted by the realities of outer and inner life, it found itself forced into the compromises and accommodations that we are all familiar with from the history of Christianity.

Nowadays, however, we have the religion of "everyday life." The numerous gods of yore, divested of their magic and hence assuming the shape of impersonal forces, arise from their graves, strive for power over our lives, and resume their eternal struggle among themselves. But what is so hard for us today, and is hardest of all for the young generation, is to meet the challenge of such an *everyday life*. All chasing after "experience" arises from this weakness. For weakness it is to be unable to look the fate of the age full in the face.

The destiny of our culture, however, is that we shall once again become more clearly conscious of this situation after a millennium in which our allegedly or supposedly exclusive reliance on the glorious pathos of the Christian ethic had blinded us to it.

But enough of these questions that lead us very far afield. For a proportion of our young people would commit a significant error here if they were to respond to all this by saying, "Very well, but the reason we come to lectures is to experience something more than just analyses and statements of fact." The error they are guilty of is that they look to the professor to be something other than he is: they are looking for a *leader* and not a *teacher*. But we are put in front of a class only as *teachers*. These are two different things and we can easily convince ourselves that this is so.

Allow me to take you back to America because it is often possible there to see things in their most basic form. An American boy learns far less than a German boy. Despite the incredible number of examinations he is subjected to, he has not yet become, as far as the *meaning* of his school life is concerned, the sort of person who is absolutely dominated by examinations that we find in Germany. For the bureaucracy that uses the examination certificate as an entry ticket to the rewards of office is still in its infancy there. The young American has no respect for anyone or anything, or any office, unless it is the personal achievement of the person concerned. *That* is what the American calls democracy. However distorted the reality may be when compared with this conception of it, it is the conception that counts here. The teacher he sees before him is someone of whom he thinks: this man sells his knowledge and grasp of method for my father's money, just as the woman in the greengrocer's sells cabbage to my mother. And that's the long and the short of it. Admittedly, if the teacher happens to be a soccer star, then he will be regarded as a leader on the soccer field. But if he is not (or has no comparable sporting achievement to his credit), he is a teacher and nothing more, and no young American would dream of letting such a teacher sell him any "worldviews" or rules for the conduct of his life. Now, put like this, we in Germany would reject such ideas. I have deliberately exaggerated here, but we may ask whether this attitude does not after all contain a grain of truth.

Fellow students! You come to our lectures with the expectation that we will be leaders, but you do not say to yourselves beforehand that out of one hundred professors, at least ninety-nine are not only not soccer stars in real life, but neither claim, nor have any right to claim, to be "leaders" of any kind in matters of conduct. Bear in mind that the value of a human being does not depend on whether he has leadership qualities. And in any case, the qualities that make someone an outstanding scholar and academic teacher are not those that create leaders in practical life or, more specifically, in politics. It is pure chance if a lecturer also has these qualities, and it would be very questionable if everyone who stands at the lectern were to feel called upon to claim them for himself. And even more questionable if it were left to every university teacher to act the leader in the lecture room. For the very people who think themselves called upon to be leaders are frequently the least qualified to be so. And, above all, whether they are leaders or not, the situation in the lecture room gives them absolutely no scope for *demonstrating their abilities*. Let the professor who feels himself called upon to advise young people and who enjoys their confidence show what he is made of in his personal relations with students, individually. And if he feels he has a vocation to intervene in the conflict of worldviews and party opinions, let him do so outside in the marketplace of life, in the press, at public meetings, in associations, or wherever he wishes. But it is all too easy for him to display the courage of his convictions in the presence of people who are condemned to silence even though they may well think differently from him.

But if all this is true, you will certainly want to ask what can science achieve positively for our "lives" at a personal and practical level? And this brings us back to the problem of its "vocation." In the first place, of course, science gives us knowledge of the techniques whereby we can control life—both external objects and human actions—through calculation. But, you will say, that is just the situation of the American boy and the woman serving in the greengrocer's. I agree entirely. But second, and this is something the greengrocer's assistant cannot do, science provides methods of thought, the tools of the trade, and the training needed to make use of them. You will perhaps object that this is not vegetables, but equally it is no more than the means by which to procure vegetables. Good, let us leave the matter open for today.

But fortunately, this is not the last word about the achievement of science, and we are in a position to offer you a third contribution, namely, *clarity*. Always assuming that clarity is something we ourselves possess. Insofar as we do, we can make clear to you that in practice we can adopt this or that attitude toward the value problem at issue—I would ask you for simplicity's sake to take examples from social phenomena. *If* you take up this or that attitude, the lessons of science are that you must apply such and such *means* in order to convert your beliefs into a reality. These means may well turn out to be of a kind that you feel compelled to reject. You will then be forced to choose between the end and the inevitable means. Does the end "justify" these means or not? The teacher can demonstrate to you the necessity of this choice. As long as he wishes to remain a teacher, and not turn into a demagogue, he can do no more. Of course, he can say to you that if you wish to achieve this or that end, you will have to put up with certain accompanying consequences that experience tells us are bound to make their appearance. So we are back to the same situation. However, these are all problems that can arise for every technician who will frequently find himself having to choose according to the principle of the lesser evil or what is relatively speaking the best option. Only in his case one principal thing is given, namely, the *end*. And it is precisely this end that is *absent* from our situation as soon as we begin to concern ourselves with "ultimate" questions.

This brings us to the last contribution that science can make in the service of clarity, and at the same time we reach its limits. We can and should tell you that the *meaning* of this or that practical stance can be inferred consistently, and hence also honestly, from this or that ultimate fundamental ideological position. It may be deducible from one position, or from a number—but there are other quite specific philosophies from which it cannot be inferred. To put it metaphorically, if you choose this particular standpoint, you will be serving this particular god and will *give offense to every other god*. For you will necessarily arrive at such-and-such ultimate, internally meaningful *conclusions* if you remain true to yourselves. We may assert this at least in principle. The discipline of philosophy and the discussion of what are ultimately the philosophical bases of the individual disciplines all attempt to achieve this. If we understand the matter correctly (something that must be assumed here) we can compel a person, or at least help him, *to render an account of the ultimate meaning of his own actions*. This seems to me to be no small matter, and can be applied to questions concerning one's own personal life. And if a teacher succeeds in this respect I would be tempted to say that he is acting in the service of "ethical" forces, that is to say, of the duty to foster clarity and a sense of responsibility. I believe that he will be all the more able to achieve this, the more scrupulously he avoids seeking to suggest a particular point of view to his listeners or even impose one on them.

The assumption that I am offering you here is based on a fundamental fact. This is that as long as life is left to itself and is understood in its own terms, it knows only that the conflict between these gods is never-ending. Or, in nonfigurative language, life is about the incompatibility of ultimate *possible* attitudes and hence the inability ever to resolve the conflicts between them. Hence the necessity of *deciding* between them. Whether in these circumstances it is worth anyone's while to choose science as a "vocation" and whether science itself has an objectively worthwhile "vocation" is itself a value judgment about which nothing useful can be said in the lecture room. This is because positively affirming the value of science is the *precondition* of all teaching. I personally answer this question in the affirmative through the very fact of my own work. And moreover, I do so on behalf of the point of view that hates intellectuality as if it were the very devil, a standpoint that modern youth endorses as its own, or at least thinks it does. For we may legitimately say to them [with Goethe], "Reflect, the Devil is old, so become old if you would understand him."[27] That is not meant literally in terms of a birth certificate, but in the sense that if you wish to get the better of this devil, there is no point in running away from him, as so often happens nowadays. Instead, you have to acquire a thorough knowledge of him so as to discover his power and his limitations.

Science today is a profession practiced in specialist *disciplines* in the service of reflection on the self and the knowledge of relationships between facts and not a gift of grace on the part of seers and prophets dispensing sacred goods and revelations. Nor is it part of the meditations of sages and philosophers about the *meaning* of the world. This is of course an ineluctable fact of our historical situation, one from which there is no escape if we remain true to ourselves. And suppose that Tolstoy rises up in you once more and asks, "who if not science will answer the question: what then shall we do and how shall we organize our lives?" Or, to put it in the language we have been using here: "Which of the warring gods shall we serve? Or shall we serve a completely different one, and who might that be?" In that event, we must reply: only a prophet or a savior. And if there is none or if his gospel is no longer believed, you will certainly not be able to force him to appear on earth by having thousands of professors appear in the guise of privileged or state-employed petty prophets and try to claim his role for themselves in their lecture rooms. If you attempt it, the only thing you will achieve will be that knowledge of a certain crucial fact will never be brought home to the younger generation in its full significance. This fact is that the prophet for whom so many of them yearn simply does *not* exist. I believe that the inner needs of a human being with the "music" of religion in his veins will never be served if the fundamental fact that his fate is to live in an age alien to God and bereft of prophets is hidden from him and others by surrogates in the shape of all these professorial prophets. The integrity of his religious sensibility must surely rise up in rebellion against this.

Now, you will be tempted to ask what we are to make of the fact that there is such a thing as "theology" and of its claims to be a "science." Let us not mince our words. "Theology" and "dogmas" are not indeed universal, but they are by no means confined to Christianity. They exist also in a highly developed form (looking back chronologically) in Islam, Manichaeism, Gnosticism, Orphism, Zoroastrianism, Buddhism, the Hindu sects, Taoism, and the Upanishads, and, of course, in Judaism. To be sure, they vary greatly in the extent to which they have been developed systematically. And in contrast to what Judaism, for example, has to show, it is no accident that Western Christianity has not only extended theology more systematically, or has striven to, but that its development has had incomparably greater historical significance. It was the Greek spirit that produced this effect, and all the theology of the West can be traced back to Greece, just as all theology of the East (obviously) goes back to Indian thought.

All theology is the intellectual *rationalization* of sacred religious beliefs. No science is absolutely free of assumptions and none can satisfactorily explain its value to a person who rejects them. But every theology adds a few assumptions that it requires for its work and thus for the justification of its existence. Their meaning and scope vary. We may say that every theology, including that of Hinduism, is based on the assumption that the world must have a *meaning*. They go on to ask how we are to interpret this meaning so that it is intellectually conceivable. The position is similar to Kant's epistemology, which proceeded from the assumption that "scientific truth exists and it is *valid*" and then went on to inquire what intellectual assumptions are required for this to be (meaningfully) possible.[28] Or as modern aesthetic philosophers (explicitly, as with Georg von Lukács, or implicitly) proceed from the assumption that "works of art *exist*" and then go on to ask how that is (meaningfully) possible.[29] Admittedly, the theologians do not content themselves as a rule with that assumption (which really belongs to the philosophy of religion). They normally proceed from a further postulate, namely, that specific "revelations" are facts vital for salvation, that is to say, facts without which the meaningful conduct of life is not possible. Therefore, these revelations simply must be believed in. Furthermore, they require you to accept that certain conditions and actions possess the quality of holiness, that is, they supply the basis or at least the elements of a life that is religiously meaningful. They then go on to ask yet again: How can these simply indispensable assumptions be meaningfully interpreted within a view of the universe as a whole? Note that for theology these assumptions lie outside the realm of "science." They are not "knowledge" in the sense ordinarily understood, but a form of "having." Whoever does not "have" them—faith or the other requisites of holiness—will not be able to obtain them with the help of theology, let alone any other branch of science. On the contrary, in every "positive" theology the believer reaches the point where St. Augustine's assertion holds good: "Credo non quod, sed *quia* absurdum est."[30] The talent for this virtuoso achievement of "sacrificing the intellect" is a crucial characteristic of men with positive religion. And the fact that this is so shows that despite (or rather as a result of) the theology (that after all reveals this fact) the tension between the value spheres of "science" and religious salvation cannot be overcome.

Properly speaking, it is only the disciple who makes a sacrifice of the intellect to the prophet, and the believer to the church. But never has a new prophecy come into being because (and I deliberately repeat a metaphor that some have found offensive) many modern intellectuals experience the need to furnish their souls, as it were, with antique objects that have been guaranteed genuine. They then recollect that religion once belonged among these antiques. It is something they do not happen to possess, but by way of a substitute they are ready to play at decorating a private chapel with pictures of the saints that they have picked up in all sorts of places, or to create a surrogate by collecting experiences of all kinds that they endow with the dignity of a mystical sanctity—and which they then hawk around the book markets. This is simply fraud or self-deception. A different phenomenon, on the other hand, is no fraud but very serious and genuine, although sometimes open to self-misinterpretation. This occurs when some of the youth organizations that have quietly grown up during recent years interpret their own human communities in religious, cosmic, or mystical terms. It may well be true that every genuinely fraternal act can be combined with the belief that it contributes something of enduring value to a suprapersonal realm. However, I think it doubtful that such religious interpretations do anything to enhance the worth of purely human relationships. But no more of that here.

Our age is characterized by rationalization and intellectualization, and above all, by the disenchantment of the world. Its resulting fate is that precisely the ultimate and most sublime values have withdrawn from public life. They have retreated either into the abstract realm of mystical life or into the fraternal feelings of personal relations between individuals. It is no accident that our greatest art is intimate rather than monumental. Nor is it a matter of chance that today it is only in the smallest groups, between individual human beings, pianissimo, that you find the pulsing beat that in bygone days heralded the prophetic spirit that swept through great communities like a firestorm and welded them together. If we attempt artificially to "invent" a sense of monumental art, this leads only to wretched monstrosities of the kind we have seen in the many artistic works of the last twenty years.

If we attempt to construct new religious movements without a new, authentic prophecy, this only gives rise to something equally monstrous in terms of inner experience, which can only have an even more dire effect. And academic prophecies can only ever produce fanatical sects, but never a genuine community. To anyone who is unable to endure the fate of the age like a man we must say that he should return to the welcoming and merciful embrace of the old churches—simply, silently, and without any of the usual public bluster of the renegade. They will surely not make it hard for him. In the process, he will inevitably be forced to make a "sacrifice of the intellect," one way or the other. We shall not bear him a grudge if he can really do it. For such a sacrifice of the intellect in favor of an unconditional religious commitment is one thing.

But morally, it is a very different thing if one shirks his straightforward duty to preserve his intellectual integrity. This is what happens when he lacks the courage to make up his mind about his ultimate standpoint but instead resorts to feeble equivocation in order to make his duty less onerous. And that embracing of religion also ranks higher to my mind than the professorial prophecy that forgets that the only morality that exists in a lecture room is that of plain intellectual integrity. This integrity enjoins us to be mindful that for all those multitudes today who are waiting for new prophets and saviors, the situation is the same as we can hear from that beautiful song of the Edomite watchman during the exile that was included in the book of Isaiah. "One calleth to me out of Seir, Watchman, what of the night? what of the night? The watchman said, Even if the morning cometh, it is still night: if ye inquire already, ye will come again and inquire once more."[31]

The people to whom this was said have inquired and waited for much longer than two thousand years, and we are familiar with its deeply distressing fate. From it we should draw the moral that longing and waiting is not enough and that we must act differently. We must go about our work and meet "the challenges of the day"—both in our human relations and our vocation.[32] But that moral is simple and straightforward if each person finds and obeys the daemon[33] that holds the threads of *his* life.

Reconstruction as Seen Twenty-Five Years Later

John Dewey

John Dewey (1859–1952) was an American philosopher and educational reformer who lived through a century of great transformations in American political and moral life. His philosophy is pragmatic in the sense of being attuned to the everyday experiences of breakdown and discord. He held that the task of thinking was to help us get clear about the character of those breakdowns and discords in a way that might open the way to practical action. In this spirit, his work rejects the search for fixed principles or timeless values in political and moral reasoning and intervention. Rather, Dewey argues that moral and political activity needs to be "reconstructed." The essay that follows is the second introduction to his book *Reconstruction in Philosophy*, written in the wake of the Second World War. In the essay Dewey urgently calls for a mode of philosophical inquiry capable of giving conceptual form to breakdowns in experience, thereby allowing us to reflect on possible means of addressing those breakdowns and moving beyond them. Inquiry in this pragmatic sense draws ethics and science into a shared practice and mutual purpose.

I

The text of this volume was written some twenty-five years ago—that is, soon after the First World War; that text is printed without revision. This Introduction is written in the spirit of the text. It is also written in the firm belief that the events of the intervening years have created a situation in which the need for reconstruction is vastly more urgent than when the book was composed; and, more specifically, in the conviction that the present situation indicates with greatly increased clearness where the needed reconstruction must center, the locus from which detailed new developments must proceed. Today Reconstruction *of* Philosophy is a more suitable title than Reconstruction *in* Philosophy. For the intervening events have sharply defined, have brought to a head, the basic postulate of the text: namely, that the distinctive office, problems and subject matter of philosophy grow out of stresses and strains in the community life in which a given form of philosophy arises, and that, accordingly, its specific problems vary with the changes in human life that are always going on and that at times constitute a crisis and a turning point in human history.

The First World War was a decided shock to the earlier period of optimism, in which there prevailed widespread belief in continued progress toward mutual understanding among peoples and classes, and hence a sure movement to harmony and peace. Today the shock is almost incredibly greater. Insecurity and strife are so general that the prevailing attitude is one of anxious and pessimistic uncertainty. Uncertainty as to what the future has in store casts its heavy and black shadow over all aspects of the present.

In philosophy today there are not many who exhibit confidence about its ability to deal competently with the serious issues of the day. Lack of confidence is manifested in concern for the improvement of techniques, and in threshing over the systems of the past. Both of these interests are justifiable in a way. But with respect to the first, the way of reconstruction is not through giving attention to form at the expense of substantial content, as is the case with techniques that are used only to develop and refine still more purely formal skills. With respect to the second, the way is not through increase of erudite scholarship about the past that throws no light upon the issues now troubling mankind. It is not too much to say that, as far as interest in the two topics just mentioned predominates, the withdrawal from the present scene, increasingly evident in philosophy, is itself a sign of the extent of the disturbance and unsettlement that now marks the other aspects of man's life. Indeed, we may go farther and say that such withdrawal is one manifestation of just those defects of past systems that render them of little value for the troubled affairs of the present: namely, the desire to find something so fixed and certain as to provide a secure refuge. The problems with which a philosophy relevant to the present must deal are those growing out of changes going on with ever-increasing rapidity, over an ever-increasing human-geographical range, and with ever-deepening intensity of penetration; this fact is one striking indication of the need for a very different kind of reconstruction from that which is now most in evidence.

When a view similar to that here presented has been advanced on previous occasions, as, indeed, in the text which follows, it has been criticized as taking what one of the milder of my critics has called "a sour attitude" toward the great systems of the past. It is, accordingly, relevant to the theme of needed reconstruction to say that the adverse criticisms of philosophies of the past are not directed at these systems with respect to their connection with intellectual and moral issues of their own time and place, but with respect to their relevancy in a much changed human situation. The very things that made the great systems objects of esteem and admiration in their own socio-cultural contexts are in large measure the very grounds that deprive them of "actuality" in a world whose main features are different to an extent indicated by our speaking of the "scientific revolution," the "industrial revolution" and the "political revolution" of the last few hundred years. A plea for reconstruction cannot, as far as I can see, be made without giving considerable critical attention to the background within which and in regard to which reconstruction is to take place. Far from being a sign of disesteem, this critical attention is an indispensable part of interest in the development of a philosophy that will do for our time and place what the great doctrines of the past did in and for the cultural media out of which they arose.

Another criticism akin to that just discussed is that the view here taken of the work and office of philosophy rests upon a romantic exaggeration of what can be accomplished by "intelligence." If the latter word were used as a synonym for what one important school of past ages called "reason" or "pure intellect," the criticism would be more than justified. But the word names something very different from what is regarded as the highest organ or "faculty" for laying hold of ultimate truths. It is a shorthand designation for great and ever-growing methods of observation, experiment and reflective reasoning which have in a very short time revolutionized the physical and, to a considerable degree, the physiological conditions of life, but which have not as yet been worked out for application to what is itself distinctively and basically *human*. It is a newcomer even in the physical field of inquiry; as yet it hasn't developed in the various aspects of the human scene. The reconstruction to be undertaken is not that of applying "intelligence" as something readymade. It is to carry over into any inquiry into human and moral subjects the kind of method (the method of observation, theory as hypothesis, and experimental test) by which understanding of physical nature has been brought to its present pitch.

Just as theories of knowing that developed prior to the existence of scientific inquiry provide no pattern or model for a theory of knowing based upon the present actual conduct of inquiry, so the earlier systems reflect both pre-scientific views of the natural world and also the pre-technological state of industry and the pre-democratic state of politics of the period when their doctrines took form. The actual conditions of life in Greece, particularly in Athens, when classic European philosophy was formulated set up a sharp division between doing and knowing, which was generalized into a complete separation of theory and "practice." It reflected, at the time, the economic organization in which "useful" work was done for the most part by slaves, leaving free men relieved from labor and "free" on that account. That such a state of affairs is also pre-democratic is clear. In political matters, nevertheless, philosophers retained the separation of theory and practice long after tools and processes derived from industrial operations had become indispensable resources in conducting the observations and experiments that are the heart of scientific knowing.

It should be reasonably obvious that an important aspect of the reconstruction that now needs to be carried out concerns the theory of knowledge. In it a radical change is demanded as to the subject matter upon which that theory must be based; the new theory will consider how knowing (that is, inquiry that is competent) is carried on, instead of supposing that it must be made to conform to views independently formed regarding faculties of organs. And, while substitution of "intelligence," in the sense just indicated, for "reason" is an important element in the change demanded, reconstruction is not confined to that matter. For the so-called empirical theories of knowledge, though they rejected the position of the rationalist school, operated in terms of what *they* took to be a necessary and sufficient faculty of knowledge, accommodating the theory of knowing to their preformed beliefs about "sense-perception" instead of deriving their view of sense-perception from what goes on in the conduct of scientific inquiry.[1]

It will be noted that the adverse criticisms dealt with in the foregoing paragraphs are dealt with not for the sake of replying to criticisms, but primarily as illustrations of why reconstruction is urgently required, and secondarily as illustrations of where it is needed. For there is no promise of the rise and growth of a philosophy relevant to the conditions that *now* supply the materials of philosophical issues and problems, save as the work of reconstruction takes serious account of how and where systems of the past indicate the need for reconstruction in the present.

II

It has been stated that philosophy grows out of, and in intention is connected with, human affairs. There is implicit in this view the further view that, while acknowledgment of this fact is a precondition of the reconstruction now required, yet it means more than that philosophy *ought* in the future to be connected with the crises and tensions in the conduct of human affairs. For it is held that in effect, if not in profession, the great systems of Western philosophy all have been thus motivated and occupied. A claim that they always have been sufficiently aware of what they were engaged in would, of course, be absurd. They have seen themselves, and have represented themselves to the public, as dealing with something which has variously been termed Being, Nature or the Universe, the Cosmos at large, Reality, the Truth. Whatever names were used they had one thing in common: they were used to designate something taken to be fixed, immutable, and therefore out of time; that is, eternal. In being also something conceived to be universal or all-inclusive, this eternal being was taken to be above and beyond all variations in space. In this matter, philosophers reflected in generalized form the popular beliefs which were current when events were thought of as taking place *in* space and time as their all-comprehensive envelopes. It is a familiar fact that the men who initiated the revolution in natural science held that space and time were independent of each other and of the things that exist and the events that take place within them. Since the assumption of underlying fixities—of which the matter of space and time and of immutable atoms is an exemplification—dominated "natural" science, there is no ground for surprise that in a more generalized form it was the foundation upon which philosophy assumed, as a matter of course, that it must erect its structure. Philosophical doctrines which disagreed about virtually everything else were at one in the assumption that their distinctive concern as philosophy was to search for the immutable and ultimate—that which *is*—without respect to the temporal or spatial. Into this state of affairs in natural science as well as in moral standards and principles, there recently entered the discovery that natural science is forced by its own development to abandon the assumption of fixity and to recognize that what for it is actually "universal" is *process*; but this fact of recent science still remains in philosophy, as in popular opinion up to the present time, a technical matter rather than what it is: namely, the most revolutionary discovery yet made.

The supposed fact that morals demand immutable, extra-temporal principles, standards, norms, ends, as the only assured protection against moral chaos can, however, no longer appeal to natural science for its support, nor expect to justify by science its exemption of morals (in practice and in theory) from considerations of time and place—that is, from processes of change. Emotional—or sentimental—reaction will doubtless continue to resist acknowledgment of this fact and refuse to use in morals the standpoint and outlook which have now made their way into natural science. But in any case, science and traditional morals have been at complete odds with one another as to the kinds of things which, according to one and the other, are immutable. Hence a deep and impassable gulf is set up between the *natural* subject matter of science and the *extra-* if not *supra*-natural subject matter of morals. There must be many thoughtful persons who are so dismayed by the inevitable consequences of this split that they will welcome that change in point of view which will render the methods and conclusions of natural science serviceable for moral theory and practice. All that is needed is acceptance of the view that moral subject matter is also spatially and temporally qualified. Considering the controverted present state of morals and its loss of popular esteem, the sacrifice demanded should not seem threatening to those who are not moved by vested institutional interest. As for philosophy, its profession of operating on the basis of the eternal and the immutable is what commits it to a function and a subject matter which, more than anything else, are the source of the growing popular disesteem and distrust of its pretensions; for it operates under cover of what is now repudiated in science, and with effective support only from old institutions whose prestige, influence and emoluments of power depend upon the preservation of the old order; and this at the very time when human conditions are so disturbed and unsettled as to call more urgently than at any previous time for the kind of comprehensive and "objective" survey in which historic philosophies have engaged. To the vested interests, maintenance of belief in the transcendence of space and time, and hence the derogation of what is "merely" human, is an indispensable prerequisite of their retention of an authority which in practice is translated into power to regulate human affairs throughout—from top to bottom.

There is, however, such a thing as relative—that is *relational*—universality. The actual conditions and occasions of human life differ widely with respect to their comprehensiveness in range and in depth of penetration. To see why such is the case, one does not have to depend upon a scientifically exploded theory of control from outside and above by self-moved and self-moving forces. On the contrary, theory began to count in the sciences of astronomy, physics, physiology, in their multiple and varied aspects, when this attitude of dogmatism was replaced by the use of hypotheses in conducting experimental observations to bind concrete facts together in systems of increasing temporal-spatial extent. The *universality* that belongs to scientific theories is not that of inherent content fixed by God or Nature, but of range of applicability—of capacity to take events out of their apparent isolation so as to order them into systems which (as is the case with all living things) prove they are alive by the kind of change which is *growth*. From the standpoint of scientific inquiry nothing is more fatal to its right to obtain acceptance than a claim that its conclusions are final and hence incapable of a development that is other than mere quantitative extension.

While I was engaged in writing this Introduction, I received a copy of an address recently delivered by a distinguished English man of science. Speaking specifically of science, he remarked, "Scientific discovery is often carelessly looked upon as the creation of some new knowledge which can be added to the great body of old knowledge. This is true of the strictly trivial discoveries. It is not true of the fundamental discoveries, such as those of the laws of mechanics, of chemical combination, of evolution, on which scientific advance ultimately depends. These always entail the destruction of or disintegration of old knowledge *before the new can be created*."[2] He continued by pointing out specific instances of the importance of getting outside of the grooves into which the heavy arm of custom tends to push every form of human activity, not excluding intellectual and scientific inquiry: "It is no accident that bacteria were first understood by a canal engineer, that oxygen was isolated by a Unitarian minister, that the theory of infection was established by a chemist, the theory of heredity by a monastic school teacher, and the theory of evolution by a man who was unfitted to be a university instructor in either botany or zoology." He closed by saying, "We need a Ministry of Disturbance, a regulated source of annoyance; a destroyer of routine; an underminer of complacency." The routine of custom tends to deaden even scientific inquiry; it stands in the way of *discovery* and of the *active* scientific worker. For discovery and inquiry are synonymous as an occupation. Science is a *pursuit*, not a coming into possession of the immutable; new theories as points of view are more prized than discoveries that quantitatively increase the store on hand. It is relevant to the theme of domination by custom that the lecturer said the great innovators in science "are the first to fear and doubt their discoveries."

I am here specially concerned with the bearing of what was said about men of science upon the work of philosophy. The borderline between what is called hypothesis in science and what is called speculation (usually in a tone of disparagement) in philosophy is thin and shadowy at the time of initiation of new movements—those placed in contrast with "technical applications and developments" such as take place as a matter of course after a new and revolutionary outlook has managed to win acceptance. Viewed in their own cultural contexts, the "hypotheses" advanced by those who now near the name of great philosophers differs from the "speculations of the men who have made great (and "destructive") innovations in science by having a wider range of reference and possible application; by the fact that they claim not to be "technical" but deeply and broadly human. At the time there is no sure way of telling whether the new way of seeing and of treating things is to turn out to be a case of science or of philosophy. Later, the classification is usually made with comparative ease. It is a case of "science" if and when its field of application is so specific, so limited, that passage into it is comparatively direct—in spite of the emotional uproar attending its appearance—as, for example, in the case of Darwin's theory. It is designated "philosophy" when its area of application is so comprehensive that it is not possible for it to pass directly into formulations of such form and content as to be serviceable in immediate conduct of specific inquiry. This fact does not signify its futility; on the contrary, the contemporary state of cultural conditions was such as to stand effectually in the way of the development of hypotheses that would give immediate direction to specific observations and experiments so definitely factual as to constitute "science." As the history of scientific inquiry clearly shows, it was during the "modern" period that inquiry took the form of discussion, which, however, was not useless or idle, scientifically speaking. For, as the word etymologically implies, this discussion was a shaking up, a stirring, which loosened the firm hold of earlier cosmology upon science. This period of discussion, with the loosening that attended it, marks the time of the shading off of what now ranks as "philosophy" into what has now attained the rank of "science."[3] What is called the "climate of opinion" is more than a matter of opinions; it is a matter of cultural habits that determine intellectual as well as emotional and volitional attitudes. The work done by the men whose names now appear in histories of philosophy rather than of science played a large role in producing a climate that was favorable to initiation of the scientific movement whose outcome is the astronomy and physics that have displaced the old ontological cosmology.

It does not need deep scholarship to be aware that, at the time, this new science was regarded as a deliberate assault upon religion and upon the morals then intimately tied up with the religion of Western Europe. Similar attacks followed the revolution that began in the nineteenth century in biology. Historical facts prove that discussions that have not been carried, because of their very comprehensive and penetrating scope, to the point of detail characteristic of science, have done a work without which science would not be what it now is.

III

The point of the foregoing discussion does not lie, however, in its bearing upon the value of past philosophic doctrines. Its relevancy for this Introduction consists of its bearing upon the reconstruction of work and subject matter that is needed to give philosophy today the vitality once possessed by its predecessors. What took place in the earlier history of science was serious enough to be named the "warfare of science and religion." Nevertheless, the scope of the events that bear that name is limited, almost technical, when it is placed in comparison with what is going on now because of the entry of science more generally into life. The present reach and thrust of what originates as science affects disturbingly every aspect of contemporary life, from the state of the family and the position of women and children, through the conduct and problems of education, through the fine as well as the industrial arts, into political and economic relations of association that are national and international in scope. They are so varied, so multiple, as well as developing with such rapidity, that they do not lend themselves to generalized statement. Moreover, their occurrence presents so many and such serious practical issues demanding immediate attention that man has been kept too busy meeting them piecemeal to make a generalized or intellectual observation of them. They came upon us like a thief in the night, taking us unawares.

The primary requisite of reconstruction is accordingly to arrive at an hypothesis as to how this great change came about so widely, so deeply, and so rapidly. The hypothesis here offered is that the upsets which, taken together, constitute the crisis in which man is now involved all over the world, in all aspects of his life, are due to the entrance into the conduct of the everyday affairs of life of processes, materials and interests whose origin lies in the work done by physical inquirers in the relatively aloof and remote technical workshops known as laboratories. It is no longer a matter of disturbance of religious beliefs and practices, but of every institution established before the rise of modern science a few short centuries ago. The earlier "warfare" was ended not by an out-and-out victory of either of the contestants but by a compromise taking the form of a division of fields and jurisdictions. In moral and ideal matters supremacy was accorded to the old. They remained virtually immutable in their older form. As the uses of the new science proved beneficial in many practical affairs, the new physical and physiological science was tolerated with the understanding that it dealt only with lower material concerns and refrained from entering the higher spiritual "realm" of Being. This "settlement" by the device of division gave rise to the dualisms which have been the chief concern of "modern" philosophy. In the developments which have actually occurred and which have culminated especially within the last generation, the settlement by division of territories and jurisdictions has completely broken down in practice. This fact is exhibited in the present vigorous and aggressive campaign of those who accept the division between the "material" and the "spiritual" but who also hold that the representatives of natural science have not stayed where they belong but have usurped in actual practice—and oftentimes in theory—the right to determine the attitudes and procedures proper to the "higher" authority. Hence, according to them, the present scene of disorder, insecurity and uncertainty, with the strife and anxiety that inevitably results.

I am not here concerned to argue directly against this view. Indeed, it may even be welcomed provided it is taken as an indication of where the issue centers with respect to reconstruction in philosophy. For it indicates by contrast the only direction which, under existing conditions, is intellectually and morally open. The net conclusion of those who hold natural science to be the *fons et origo* of the undeniably serious ills of the present is the necessity of bringing science under subjection to some special institutional "authority." The alternative is a generalized reconstruction so fundamental that it has to be developed by recognition that while the evils resulting at present from the entrance of "science" into our common ways of living are undeniable they are due to the fact that no systematic efforts have as yet been made to subject the "morals" underlying old institutional customs to scientific inquiry and criticism. Here, then, lies the reconstructive work to be done by philosophy. It must undertake to do for the development of inquiry into human affairs and hence into morals what the philosophers of the last few centuries did for promotion of scientific inquiry in physical and physiological conditions and aspects of human life.

This view of what philosophy needs in order to be relevant to present human affairs and to regain the vitality it is losing is not concerned to deny that the entry of science into human activities and interests has its destructive phase. Indeed, the point of departure for the view here presented regarding the reconstruction demanded in philosophy is that this entry, amounting to a hostile invasion of the old, is the main factor operating to produce the present estate of man. And, while the attack upon science as the responsible and guilty party is terribly one-sided in its emphasis upon the destruction involved and in neglect of the many and great human benefits that have accrued, it is held that the issue cannot be disposed of by drawing a balance sheet of human loss and gain with a view to showing that the latter predominates.

The case in fact is much simpler. The premise on which the present assault upon science depends is that old institutional customs, including institutional belief, provide an adequate, and indeed a final, criterion by which to judge the worth of consequences produced by the disturbing entry of science. Those who maintain this premise systematically refuse to note that "science" has a copartner in producing our critical situation. It only takes an eye single to the facts to observe that science, instead of operating alone and in a void, works within an institutional state of affairs developed in prescientific days, one which is not modified by scientific inquiry into the moral principles that were then formed and were, presumably, appropriate to it.

One simple example shows the defection and distortion that results from viewing science in isolation. The destructive use made of the fission of the nucleus of an atom has become the stock-in-trade of the assault upon science. What is so ignored as to be denied is that this destructive consequence occurred not only in a war but because of the existence of war, and that war as an institution antedates by unknown millennia the appearance on the human scene of anything remotely resembling scientific inquiry. That in this case destructive consequences are directly due to pre-existent institutional conditions is too obvious to call for argument. It does not prove that such is the case everywhere and at all times; but it certainly cautions us against the irresponsible and indiscriminate dogmatism now current. It gives us the definite advice to recall the unscientific conditions under which morals, in both the practical and the theoretical senses of that word, took on form and content. The end-in-view in calling attention to a fact that cannot be denied, but that is systematically ignored, is not the futile, because totally irrelevant, purpose of justifying the work of scientific inquirers in general or in special cases. It is to direct attention to a fact of outstanding intellectual import. The development of scientific inquiry is immature; it has not as yet got beyond the physical and physiological aspects of human concerns, interests and subject matters. In consequence, it has partial and exaggerated effects. *The institutional conditions into which it enters and which determine its human consequences have not as yet been subjected to any serious, systematic inquiry worthy of being designated scientific.*

The bearing of this state of affairs upon the present state of philosophy and the reconstruction which should be undertaken is the theme and thesis of this Introduction. Before directly resuming that theme, I shall say something about the present state of morals: a word, be it remembered, that stands both for a morality as a practical socio-cultural fact in respect to matters of right and wrong, good and evil, and for theories about the ends, standards, principles according to which the actual state of affairs is to be surveyed and judged. Now the simple fact of the case is that any inquiry into what is deeply and inclusively human enters perforce into the specific area of morals. It does so whether it intends to and whether it is even aware of it or not. When "sociological" theory withdraws from consideration of the basic interests, concerns, the actively moving aims, of a human culture on the ground that "values" are involved and that inquiry as "scientific" has nothing to do with values, the inevitable consequence is that inquiry in the human area is confined to what is superficial and comparatively trivial, no matter what its parade of technical skills. But, on the other hand, if and when inquiry attempts to enter in critical fashion into that which is human in its full sense, it comes up against the body of prejudices, traditions and institutional customs that consolidated and hardened in a pre-scientific age. For it is tautology, not the announcement of a discovery or of an inference, to state that morals, in both senses of the word, are pre-scientific when formed in an age preceding the rise of science as now understood and practiced. And to be unscientific, when human affairs in the concrete are immensely altered, is in effect to resist the formation of methods of inquiry into morals in a way that renders existing morals—again in both senses—anti-scientific.

The case would be comparatively simple if there were already in hand the intellectual standpoint, outlook, or what philosophy has called "categories," to serve as instrumentalities of inquiry. But to assume that they are at hand is to assume that intellectual growths which reflect a pre-scientific state of human affairs, concerns, interests and ends are adequate to deal with a human situation which is increasingly and for a very large part the outgrowth of new science. In a word, it is to decide to continue the present state of drift, instability and uncertainty. If the foregoing statements are understood in the sense in which they are intended, the view that is here proposed in regard to reconstruction in philosophy will stand out forcibly. From the position here taken, reconstruction can be nothing less than the work of developing, of forming, of producing (in the literal sense of that word) the intellectual instrumentalities which will progressively direct inquiry into the deeply and inclusively human—that is to say, moral—facts of the present scene and situation.

The first step, a prerequisite of further steps in the same general direction, will be to recognize that, factually speaking, the present human scene, for good and evil, for harm and benefit alike, is what it is because, as has been said, of the entry into everyday and common (in the sense of ordinary and of shared) ways of living of what has its origin in *physical* inquiry. The methods and conclusions of "science" do not remain penned in within "science." Even those who conceive of science as if it were a self-enclosed, self-actuated independent and isolated entity cannot deny that it does not remain such in practical fact. It is a piece of theoretical animistic mythology to view it as an entity, as do those who hold that it is *fons et origo* of present human woes. The science that has so far found its way deeply and widely into the actual affairs of human life is partial and incomplete science: competent in respect to physical, and now increasingly to physiological, conditions (as is seen in the recent developments in medicine and public sanitation), but nonexistent with respect to matters of supreme significance to man—those which are distinctively of, for, and by, man. No intelligent way of seeing and understanding the present estate of man will fail to note the extraordinary split in life occasioned by the radical incompatibility between operations that manifest and perpetuate the morals of a prescientific age and the operations of a scene which has suddenly, with immense acceleration and with thorough pervasiveness, been factually determined by a science which is still partial, incomplete, and of necessity one-sided in operation.

IV

In what precedes, reference has been made several times to what certain human beings classed as philosophers accomplished in the seventeenth, eighteenth and nineteenth centuries in the way of clearing the ground of cosmological and ontological debris which had been absorbed emotionally and intellectually into the very structure and operation of Western culture. It was not claimed that credit for the specific inquiries which progressively revolutionized astronomy, physics (including chemistry) and physiology belongs to philosophers. It is recorded as matter of historic fact that the latter performed an office that, given the accepted cultural climate and the momentum of accepted custom, was an indispensable prerequisite of what men of science accomplished. What will now be added to that statement, in conjunction with its bearing upon reconstruction of philosophy, is that in doing their specific jobs scientific men worked out a method of inquiry so inclusive in range and so penetrating, so pervasive and so universal, as to provide the pattern and model which permits, invites and even demands the kind of formulation that falls within the function of philosophy. It is a method of knowing that is self-corrective in operation; that learns from failures as from successes. The heart of the method is the discovery of the identity of inquiry with discovery. Within the specialized, the relatively technical, activities of natural science, this office of discovery, of uncovering the new and leaving behind the old, is taken for granted. Its similar centrality in every form of intellectual activity is, however, so far from enjoying general recognition that, in matters which are set apart as "spiritual" and "ideal" and as distinctively moral, the mere idea of it shocks many who take it as a matter of course in their own specialized work. It is a familiar fact that the practical correlate of discovery when it is scientific and theoretical is *invention*, and that in many of the physical aspects of human affairs there is even now a generalized method for the invention of inventions. In what is distinctively human, invention rarely occurs, and then only in the stress of an emergency. In human affairs and in relations that range extensively and penetrate deeply the mere idea of invention awakens fear and horror, being regarded as dangerous and destructive. This fact, which is important but which rarely receives notice, is assumed to belong to the very nature and essence of morals as morals. This fact testifies both to the reconstruction to be undertaken and to the extreme difficulty of every attempt to bring it about.

The adjustment which finally moderated, without completely exorcising, the earlier split between science and received institutional customs was a truce rather than anything remotely approaching integration. It consisted, in fact, of a device that was the exact opposite of integration. It operated on the basis of a hard and fast division of the interests, concerns and purposes of human activity into two "realms," or, by a curious use of language, into two "spheres"—not hemispheres. One was taken to be "high" and hence to possess supreme jurisdiction over the other as inherently "low." That which is high was given the name "spiritual," ideal, and was identified with the moral. The other was the "physical" as determined by the procedures of the new science of nature. In being low it was material; its methods were fitted only to the materialistic and to the world of sense-perception, not to that of reason and revelation. The new natural science was grudgingly given a license to operate on condition that it stay in its own compartment and mind its own business, as thus determined for it. That for philosophy the outcome was the whole brood and nest of dualisms which have, upon the whole, formed the "problems" of philosophy termed "modern" is a reflection of the cultural conditions which account for the basic split made between the moral and the physical. These words stand in fact for the attempt to obtain the practical advantages of ease, comfort, convenience and power that were the outcome of the "application" of the new science to the ordinary affairs of life, while retaining intact the supreme authority of the old in those matters of high morals named "spiritual." The material and utilitarian advantages of the new science, rather than anything approaching acknowledgment of the intellectual—to say nothing of the moral—import of the new method, turned out to be the most dependable ally of the men who produced the new method of revolutionizing what had been taken to be a scientific account of nature as cosmos.

The truce endured for a time. The equilibrium it presented was decidedly uneasy. The saying about keeping a cake and at the same time eating it is applicable. It represented the effort to enjoy the material and practical or utilitarian advantages of the new science while preventing its serious impact on old institutional habits—including those of belief—that were accepted as the foundation of norms and moral principles. In consequence the division would not stay put. Upon the whole, without deliberate intent (though with considerable deliberate encouragement from one group of "advanced" philosophical thinkers) the consequences issuing from the uses to which the new science was put crowded in upon the activities and values nominally reserved for the "spiritual." The impact of this encroachment constitutes what is called secularization, a movement which, as it extended itself, was regarded as a sacrilegious profaning of the sacredness of the spiritual. Even today many men who are in no way practically identified with old ecclesiastical institutions, or with the metaphysics associated with it, speak regretfully and at best apologetically of this secularization. Yet the opportunity for any genuine universalization of the method—and spirit—of science as inquiry, which is perforce discovery in which old intellectual attitudes and conclusions are unceasingly yielding to the different and new, lies precisely in discovering how to give the factors of this secularization the shape, the content and the authority nominally assigned to morals, but not now exercised in fact by those morals that have come down to us from a pre-scientific age. The actuality of this loss of authority is acknowledged in the current revival of the old doctrine of the inherent depravity of human nature to account for the loss, as well as being shown in widespread pessimism as to the future of man. These complaints and doubts are warranted as long as one regards the institutional customs in action and belief of a pre-scientific age as ultimate and immutable. But they also apply, if they are employed that way, a challenge to develop a theory of morals that will give positive intellectual direction to man in developing the practical—that is, actually effective— morals which will utilize the resources now at our disposal to bring into the activities and interests of human life order and security, not only in place of confusion but on a wider scale than ever existed in the past.

Three things are intimately connected in the plaints and promulgations that are temporarily most vocal. They are: (1) the attack upon natural science; (2) the doctrine that man is so inherently corrupt that it is impossible to form morals which will operate in behalf of stability, equity and (true) freedom without recourse to an extra-human and extra-natural authority; and (3) the claim put forth by representatives of some particular kind of institutional organization, that they alone can do what is needed. I do not mention this matter here in order to subject it to direct criticism. I mention it because it presents a position so generalized as clearly to indicate one direction in which philosophy may move out of the apathy of irrelevance. By sharp contrast, it points to the other direction in which philosophy may proceed: that of systematic endeavor to see and to state the constructive significance for the future of man issuing from the revolution wrought primarily by the new science; provided we exercise resolute wisdom in developing a system of belief-attitudes, a philosophy, framed on the basis of the resources now at our command.

The issue actually raised by the assault upon the new science and its offspring by wholesale condemnation of human nature, and by the plea to reinstate in full measure the authority of antique medieval institutions, is simply whether we are to move forward in a direction made possible by these new resources or whether the latter are so inherently untrustworthy that we must bring them under control by subjection to an authority claiming to be extra-human and extra-natural—as far as the import of "natural" is determined by scientific inquiry. The impact of systematic perception of this cleavage of directions upon philosophy is disclosure that what is called "modern" is as yet unformed, inchoate. Its confused strife and its unstable uncertainties reflect the mixture of an old and a new that are incompatible. The genuinely modern has still to be brought into existence. The work of actual production is not the task or responsibility of philosophy. That work can be done only by the resolute, patient, co-operative activities of men and women of good will, drawn from every useful calling, over an indefinitely long period. There is no absurd claim made that philosophers, scientists or any other one group form a sacred priesthood to whom the work is entrusted. But, as philosophers in the last few centuries have performed a useful and needed work in furtherance of physical inquiry, so their successors now have the opportunity and the challenge to do a similar work in forwarding moral inquiry. The conclusions of that inquiry by themselves would no more constitute a complete moral theory and a working science of distinctively human subject matter than the activities of their predecessors brought the physical and physiological conditions of human existence into direct and full-fledged existence. But it would have an active share in the work of construction of a moral human science which serves as a needful precursor of *re*construction of the actual state of human life toward order and toward other conditions of a fuller life than man has yet enjoyed.

Systematic exposure of how, where and why philosophies appropriate to ancient and medieval conditions and to those of the few centuries which have elapsed since the appearance of natural science on the human scene is so irrelevant as to be obstructive in intellectual dealings with the present scene, is itself a challenging intellectual task. As earlier intimated, reconstruction is not something to be accomplished by finding fault or being querulous. It is strictly an intellectual work demanding the widest possible scholarship as to the connections of past systems with the cultural conditions that set their problems and a knowledge of present-day science which is other than that of "popular" expositions. And this negative aspect of the intellectual activity to be performed involves of necessity a systematic exploration of the values belonging to what is genuinely new in the scientific, technological and political movements of the immediate past and of the present, when they are liberated from the incubus imposed on them by habits formed in a pre-scientific, pre-technological-industrial and pre-democratic political period. One now fairly often runs across signs of a growing tendency to react against the view which holds that science and the new technology are to be blamed for present evils. It is recognized that as means they are so powerful as to give us valuable new resources. All that is needed, so it is held, is an equally effective moral renewal that will use these means for genuinely human ends. This position is certainly a marked improvement upon a mere assault on science and technology for the purpose of effecting a specific institutional subordination of them. It is to be welcomed in so far as it perceives that the matter at issue is moral or human. But—at least in the cases in which I have met it—it suffers from a serious defect. It appears to assume that we already have in our possession, readymade, so to say, the morals that determine the ends for which the greatly enhanced store of means should be used. The *practical* difficulty in the way of rendering radically new "means" into servants of ends framed when the means at our disposal were of a different kind is ignored. But much more important than this, with respect to theory or philosophy, is the fact that it retains intact the divorce between some things as means and mere means and other things as ends and only ends because of their own essence or inherent nature. Thus in effect, though not in intent, an issue which is serious enough to be *moral* is disastrously evaded.

Just as this separation of some things as ends-in-themselves from other things as means-in-themselves, by their very nature, is a heritage of an age in which only those activities were called "useful" which served living physiologically rather than morally, and which were carried on by slaves or serfs to serve men who were *free* in the degree to which they were relieved from the need of labor that was base and material, so the primary need of the new state in which resources vastly different both qualitatively and quantitatively are at our command involves formation of new ends, ideals and standards to which to attach our new means. It is morally as well as logically impossible that a thoroughly changed kind of means should be harnessed to ends which at most are supposed to be changed only in the ease with which they can be reached. The thoroughgoing secularization of means and opportunities that has been going on has so far revolutionized the conduct of life as to have unsettled the old scene. Nothing is more intellectually futile (as well as practically impossible) than to suppose harmony and order can be achieved except as new ends and standards, new moral principles, are first developed with a reasonable degree of clarity and system.

In short, the problem of reconstruction in philosophy, from whatever angle it is approached, turns out to have its inception in the endeavor to discover how the new movements in science and in the industrial and political human conditions which have issued from it, that are as yet only inchoate and confused, shall be carried to completion. For a fulfillment which is consonant with their own, their proper, direction and momentum of movement can be achieved only in terms of ends and standards so distinctively human as to constitute a new moral order.

It is for the future to undertake, even in their philosophic aspect, the specific reconstructions that are involved in this carrying on to fulfillment what we have as yet attained only partially. Even a satisfactory listing of the issues that are involved with respect to philosophy must, by and large, wait till the philosophic movement in this direction has been carried beyond any point as yet attained. But one outstanding member of such a list has just received incidental attention: namely, the divorce that was set up between mere means and ends in themselves, which is the theoretical correlate of the sharp division of men into free and slave, superior and inferior. Science as conducted, science in practice, has completely repudiated these separations and isolations.

Scientific inquiry has raised activities, materials, tools, of the type once regarded as practical (in a low utilitarian sense) into itself; it has incorporated them into its own being. The way work is carried on in any astronomical observatory in the land, as well as in any physical laboratory, is evidence. Theory in formal statement also is as yet far behind theory in scientific practice. Theory in fact—that is, in the conduct of scientific inquiry—has lost ultimacy. Theories have passed into hypotheses. It remains for philosophy to point out in particular and in general the untold significance of this fact for morals. For in what is now taken to be morals the fixed, the immutable, still reign, even though moral theorists and moral institutional dogmatists are at complete odds with one another as to what ends, standards and principles are the ones which are immutable, eternal and universally applicable. In science the order of fixities has already passed irretrievably into an order of connections *in process*. One of the most immediate duties of philosophical reconstruction with respect to the development of viable instruments for inquiry into human or moral facts is to deal systematically with *human* processes.

Attention was earlier given in passing to some current misconceptions of the position set forth in the text which follows. I conclude with explicit notice of a point that has received repeated mention in the preceding text of the present Introduction. It has been charged that the view here taken of the work and subject matter of philosophy commits those who accept it to identification of philosophy with the work of those men called "reformers"—whether with praise or with disparagement. In a verbal sense re-form and re-construction are close together. But the re-construction or re-form here presented is strictly one of theory of the type that is so comprehensive in scope as to constitute philosophy. One of the operations to be undertaken in a re-constructed philosophy is to assemble and present reasons why the separation once set up between theory and practice no longer exists, so that a man like Justice Holmes can say that theory is the most practical thing, for good or for evil, in the world. One may hope surely that the theoretical enterprise herein presented will bear practical issue and for good. But that achievement is the work of human beings as human, not of them in any special professional capacity.

What Is Enlightenment?

Michel Foucault

Michel Foucault (1926–1984) was a philosopher and historian and one of the most original and dynamic thinkers of the postwar period. In this paper, published after his untimely death, he returns to Immanuel Kant's essay in order to repose the question of enlightenment as a central problem for the role and purpose of thinking as part of, and as a response to, modernity. Foucault lays out parameters for critical inquiry into modernity, the parameters for conducting such inquiry, and the ethical stakes of thinking as part of a modern ethos.

I

Today when a periodical asks its readers a question, it does so in order to collect opinions on some subject about which everyone has an opinion already; there is not much likelihood of learning anything new. In the eighteenth century, editors preferred to question the public on problems that did not yet have solutions. I don't know whether or not that practice was more effective; it was unquestionably more entertaining.

In any event, in line with this custom, in November 1784 a German periodical, *Berlinische Monatsschrift*, published a response to the question: *Was ist Aufklärung?* And the respondent was Kant. A minor text, perhaps. But it seems to me that it marks the discreet entrance into the history of thought of a question that modern philosophy has not been capable of answering, but that it has never managed to get rid of, either. And one that has been repeated in various forms for two centuries now. From Hegel through Nietzsche or Max Weber to Horkheimer or Habermas, hardly any philosophy has failed to confront this same question, directly or indirectly. What, then, is this event that is called the *Aufklärung* and that has determined, at least in part, what we are, what we think, and what we do today? Let us imagine that the *Berlinische Monatsschrift* still exists and that it is asking its readers the question: What is modern philosophy? Perhaps we could respond with an echo: modern philosophy is the philosophy that is attempting to answer the question raised so imprudently two centuries ago: *Was ist Aufklärung?*

Let us linger a few moments over Kant's text. It merits attention for several reasons.

1. To this same question, Moses Mendelssohn had also replied in the same journal, just two months earlier. But Kant had not seen Mendelssohn's text when he wrote his. To be sure, the encounter of the German philosophical movement with the new development of Jewish culture does not date from this precise moment. Mendelssohn had been at that crossroads for thirty years or so, in company with Lessing. But up to this point it had been a matter of making a place for Jewish culture within German thought—which Lessing had tried to do in *Die Juden*—or else of identifying problems common to Jewish thought and to German philosophy; this is what Mendelssohn had done in his *Phädon; Oder, Über die Unsterblichkeit der Seele*. With the two texts published in the *Berlinische Monatsschrift* the German *Aufklärung* and the Jewish *Haskala* recognize that they belong to the same history; they are seeking to identify the common processes from which they stem. And it is perhaps a way of announcing the acceptance of a common destiny—we now know to what drama that was to lead.

2. But there is more. In itself and within the Christian tradition, Kant's text poses a new problem. It was certainly not the first time that philosophical thought had sought to reflect on its own present. But, speaking schematically, we may say that this reflection had until then taken three main forms:

The present may be represented as belonging to a certain era of the
    world, distinct from the others through some inherent characteristics,
    or separated from the others by some dramatic event. Thus, in Plato's
    Statesman the interlocutors recognize that they belong to one of those

revolutions of the world in which the world is turning backwards, with all the negative consequences that may ensue.

The present may be interrogated in an attempt to decipher in it the heralding signs of a forthcoming event. Here we have the principle of a kind of historical hermeneutics of which Augustine might provide an example.

The present may also be analyzed as a point of transition toward the dawning of a new world. That is what Vico describes in the last chapter of *La Scienza Nuova*; what he sees "today" is "a complete humanity . . . spread abroad through all nations, for a few great monarchs rule over this world of peoples"; it is also "Europe . . . radiant with such humanity that it abounds in all the good things that make for the happiness of human life."[1]

Now the way Kant poses the question of *Aufklärung* is entirely different: it is neither a world era to which one belongs, nor an event whose signs are perceived, nor the dawning of an accomplishment. Kant defines *Aufklärung* in an almost entirely negative way, as an *Ausgang*, an "exit," a "way out." In his other texts on history, Kant occasionally raises questions of origin or defines the internal teleology of a historical process. In the text on *Aufklärung*, he deals with the question of contemporary reality alone. He is not seeking to understand the present on the basis of a totality or of a future achievement. He is looking for a difference: What difference does today introduce with respect to yesterday?

3. I shall not go into detail here concerning this text, which is not always very clear despite its brevity. I should simply like to point out three or four features that seem to me important if we are to understand how Kant raised the philosophical question of the present day.

Kant indicates right away that the "way out" that characterizes Enlightenment is a process that releases us from the status of "immaturity." And by "immaturity," he means a certain state of our will that makes us accept someone else's authority to lead us in areas where the use of reason is called for. Kant gives three examples: we are in a state of "immaturity" when a book takes the place of our understanding, when a spiritual director takes the place of our conscience, when a doctor decides for us what our diet is to be. (Let us note in passing that the register of these three critiques is easy to recognize, even though the text does not make it explicit.) In any case, Enlightenment is defined by a modification of the preexisting relation linking will, authority, and the use of reason.

We must also note that this way out is presented by Kant in a rather ambiguous manner. He characterizes it as a phenomenon, an ongoing process; but he also presents it as a task and an obligation. From the very first paragraph, he notes that man himself is responsible for his immature status. Thus it has to be supposed that he will be able to escape from it only by a change that he himself will bring about in himself. Significantly, Kant says that this Enlightenment has a *Wahlspruch*: now a *Wahlspruch* is a heraldic device, that is, a distinctive feature by which one can be recognized, and it is also a motto, an instruction that one gives oneself and proposes to others. What, then, is this instruction? *Aude sapere*: "dare to know," "have the courage, the audacity, to know." Thus Enlightenment must be considered both as a process in which men participate collectively and as an act of courage to be accomplished personally. Men are at once elements and agents of a single process. They may be actors in the process to the extent that they participate in it; and the process occurs to the extent that men decide to be its voluntary actors.

A third difficulty appears here in Kant's text in his use of the word "mankind," *Menschheit*. The importance of this word in the Kantian conception of history is well known. Are we to understand that the entire human race is caught up in the process of Enlightenment? In that case, we must imagine Enlightenment as a historical change that affects the political and social existence of all people on the face of the earth. Or are we to understand that it involves a change affecting what constitutes the humanity of human beings? But the question then arises of knowing what this change is. Here again, Kant's answer is not without a certain ambiguity. In any case, beneath its appearance of simplicity, it is rather complex.

Kant defines two essential conditions under which mankind can escape from its immaturity. And these two conditions are at once spiritual and institutional, ethical and political.

The first of these conditions is that the realm of obedience and the realm of the use of reason be clearly distinguished. Briefly characterizing the immature status, Kant invokes the familiar expression: "Don't think, just follow orders"; such is, according to him, the form in which military discipline, political power, and religious authority are usually exercised. Humanity will reach maturity when it is no longer required to obey, but when men are told: "Obey, and you will be able to reason as much as you like." We must note that the German word used here is *räsonieren*; this word, which is also used in the *Critiques* does not refer to just any use of reason, but to a use of reason in which reason has no other end but itself: *räsonieren* is to reason for reasoning's sake. And Kant gives examples, these too being perfectly trivial in appearance: paying one's taxes, while being able to argue as much as one likes about the system of taxation, would be characteristic of the mature state; or again, taking responsibility for parish service, if one is a pastor, while reasoning freely about religious dogmas.

We might think that there is nothing very different here from what has been meant, since the sixteenth century, by freedom of conscience: the right to think as one pleases so long as one obeys as one must. Yet it is here that Kant brings into play another distinction, and in a rather surprising way. The distinction he introduces is between the private and public uses of reason. But he adds at once that reason must be free in its public use, and must be submissive in its private use. Which is, term for term, the opposite of what is ordinarily called freedom of conscience.

But we must be somewhat more precise. What constitutes, for Kant, this private use of reason? In what area is it exercised? Man, Kant says, makes a private use of reason when he is "a cog in a machine"; that is, when he has a role to play in society and jobs to do: to be a soldier, to have taxes to pay, to be in charge of a parish, to be a civil servant, all this makes the human being a particular segment of society; he finds himself thereby placed in a circumscribed position, where he has to apply particular rules and pursue particular ends. Kant does not ask that people practice a blind and foolish obedience, but that they adapt the use they make of their reason to these determined circumstances; and reason must then be subjected to the particular ends in view. Thus there cannot be, here, any free use of reason.

On the other hand, when one is reasoning only in order to use one's reason, when one is reasoning as a reasonable being (and not as a cog in a machine), when one is reasoning as a member of reasonable humanity, then the use of reason must be free and public. Enlightenment is thus not merely the process by which individuals would see their own personal freedom of thought guaranteed. There is Enlightenment when the universal, the free, and the public uses of reason are superimposed on one another.

Now this leads us to a fourth question that must be put to Kant's text. We can readily see how the universal use of reason (apart from any private end) is the business of the subject himself as an individual; we can readily see, too, how the freedom of this use may be assured in a purely negative manner through the absence of any challenge to it; but how is a public use of that reason to be assured? Enlightenment, as we see, must not be conceived simply as a general process affecting all humanity; it must not be conceived only as an obligation prescribed to individuals: it now appears as a political problem. The question, in any event, is that of knowing how the use of reason can take the public form that it requires, how the audacity to know can be exercised in broad daylight, while individuals are obeying as scrupulously as possible. And Kant, in conclusion, proposes to Frederick II, in scarcely veiled terms, a sort of contract—what might be called the contract of rational despotism with free reason: the public and free use of autonomous reason will be the best guarantee of obedience, on condition, however, that the political principle that must be obeyed itself be in conformity with universal reason.

Let us leave Kant's text here. I do not by any means propose to consider it as capable of constituting an adequate description of Enlightenment; and no historian, I think, could be satisfied with it for an analysis of the social, political, and cultural transformations that occurred at the end of the eighteenth century.

Nevertheless, notwithstanding its circumstantial nature, and without intending to give it an exaggerated place in Kant's work, I believe that it is necessary to stress the connection that exists between this brief article and the three *Critiques*. Kant in fact describes Enlightenment as the moment when humanity is going to put its own reason to use, without subjecting itself to any authority; now it is precisely at this moment that the critique is necessary, since its role is that of defining the conditions under which the use of reason is legitimate in order to determine what can be known, what must be done, and what may be hoped. Illegitimate uses of reason are what give rise to dogmatism and heteronomy, along with illusion; on the other hand, it is when the legitimate use of reason has been clearly defined in its principles that its autonomy can be assured. The critique is, in a sense, the handbook of reason that has grown up in Enlightenment; and, conversely, the Enlightenment is the age of the critique.

It is also necessary, I think, to underline the relation between this text of Kant's and the other texts he devoted to history. These latter, for the most part, seek to define the internal teleology of time and the point toward which the history of humanity is moving. Now the analysis of Enlightenment, defining this history as humanity's passage to its adult status, situates contemporary reality with respect to the overall movement and its basic directions. But at the same time, it shows how, at this very moment, each individual is responsible in a certain way for that overall process.

The hypothesis I should like to propose is that this little text is located in a sense at the crossroads of critical reflection and reflection on history. It is a reflection by Kant on the contemporary status of his own enterprise. No doubt it is not the first time that a philosopher has given his reasons for undertaking his work at a particular moment. But it seems to me that it is the first time that a philosopher has connected in this way, closely and from the inside, the significance of his work with respect to knowledge, a reflection on history and a particular analysis of the specific moment at which he is writing and because of which he is writing. It is in the reflection on "today" as difference in history and as motive for a particular philosophical task that the novelty of this text appears to me to lie.

And, by looking at it in this way, it seems to me we may recognize a point of departure: the outline of what one might call the attitude of modernity.

II

I know that modernity is often spoken of as an epoch, or at least as a set of features characteristic of an epoch; situated on a calendar, it would be preceded by a more or less naive or archaic premodernity, and followed by an enigmatic and troubling "postmodernity." And then we find ourselves asking whether modernity constitutes the sequel to the Enlightenment and its development, or whether we are to see it as a rupture or a deviation with respect to the basic principles of the eighteenth century.

Thinking back on Kant's text, I wonder whether we may not envisage modernity rather as an attitude than as a period of history. And by "attitude," I mean a mode of relating to contemporary reality; a voluntary choice made by certain people; in the end, a way of thinking and feeling; a way, too, of acting and behaving that at one and the same time marks a relation of belonging and presents itself as a task. A bit, no doubt, like what the Greeks called an *ethos*. And consequently, rather than seeking to distinguish the "modern era" from the "premodern" or "postmodern," I think it would be more useful to try to find out how the attitude of modernity, ever since its formation, has found itself struggling with attitudes of "countermodernity."

To characterize briefly this attitude of modernity, I shall take an almost indispensable example, namely, Baudelaire; for his consciousness of modernity is widely recognized as one of the most acute in the nineteenth century.

1. Modernity is often characterized in terms of consciousness of the discontinuity of time: a break with tradition, a feeling of novelty, of vertigo in the face of the passing moment. And this is indeed what Baudelaire seems to be saying when he defines modernity as "the ephemeral, the fleeting, the contingent."[2] But, for him, being modern does not lie in recognizing and accepting this perpetual movement; on the contrary, it lies in adopting a certain attitude with respect to this movement; and this deliberate, difficult attitude consists in recapturing something eternal that is not beyond the present instant, nor behind it, but within it. Modernity is distinct from fashion, which does no more than call into question the course of time; modernity is the attitude that makes it possible to grasp the "heroic" aspect of the present moment. Modernity is not a phenomenon of sensitivity to the fleeting present; it is the will to "heroize" the present.

I shall restrict myself to what Baudelaire says about the painting of his contemporaries. Baudelaire makes fun of those painters who, finding nineteenth-century dress excessively ugly, want to depict nothing but ancient togas. But modernity in painting does not consist, for Baudelaire, in introducing black clothing onto the canvas. The modern painter is the one who can show the dark frock-coat as "the necessary costume of our time," the one who knows how to make manifest, in the fashion of the day, the essential, permanent, obsessive relation that our age entertains with death. "The dress-coat and frock-coat not only possess their political beauty, which is an expression of universal equality, but also their poetic beauty, which is an expression of the public soul—an immense cortège of undertaker's mutes (mutes in love, political mutes, bourgeois mutes . . .). We are each of us celebrating some funeral."[3] To designate this attitude of modernity, Baudelaire sometimes employs a litotes that is highly significant because it is presented in the form of a precept: "You have no right to despise the present."

2. This heroization is ironical, needless to say. The attitude of modernity does not treat the passing moment as sacred in order to try to maintain or perpetuate it. It certainly does not involve harvesting it as a fleeting and interesting curiosity. That would be what Baudelaire would call the spectator's posture. The *flâneur*, the idle, strolling spectator, is satisfied to keep his eyes open, to pay attention and to build up a storehouse of memories. In opposition to the *flâneur*, Baudelaire describes the man of modernity: "Away he goes, hurrying, searching. . . . Be very sure that this man . . . —this solitary, gifted with an active imagination, ceaselessly journeying across the great human desert—has an aim loftier than that of a mere *flâneur*, an aim more general, something other than the fugitive pleasure of circumstance. He is looking for that quality which you must allow me to call 'modernity.' . . . He makes it his business to extract from fashion whatever element it may contain of poetry within history." As an example of modernity, Baudelaire cites the artist Constantin Guys. In appearance a spectator, a collector of curiosities, he remains "the last to linger wherever there can be a glow of light, an echo of poetry, a quiver of life or a chord of music; wherever a passion can *pose* before him, wherever natural man and conventional man display themselves in a strange beauty, wherever the sun lights up the swift joys of the *depraved animal*."[4]

But let us make no mistake. Constantin Guys is not a *flâneur*; what makes him the modern painter par excellence in Baudelaire's eyes is that, just when the whole world is falling asleep, he begins to work, and he transfigures that world. His transfiguration does not entail an annulling of reality, but a difficult interplay between the truth of what is real and the exercise of freedom; "natural" things become "more than natural," "beautiful" things become "more than beautiful," and individual objects appear "endowed with an impulsive life like the soul of their creator."[5] For the attitude of modernity, the high value of the present is indissociable from a desperate eagerness to imagine it, to imagine it otherwise than it is, and to transform it not by destroying it but by grasping it in what it is. Baudelairean modernity is an exercise in which extreme attention to what is real is confronted with the practice of a liberty that simultaneously respects this reality and violates it.

3. However, modernity for Baudelaire is not simply a form of relationship to the present; it is also a mode of relationship that has to be established with

oneself. The deliberate attitude of modernity is tied to an indispensable asceticism. To be modern is not to accept oneself as one is in the flux of the passing moments; it is to take oneself as object of a complex and difficult elaboration: what Baudelaire, in the vocabulary of his day, calls *dandysme*. Here I shall not recall in detail the well-known passages on "vulgar, earthy, vile nature"; on man's indispensable revolt against himself; on the "doctrine of elegance" which imposes "upon its ambitious and humble disciples" a discipline more despotic than the most terrible religions; the pages, finally, on the asceticism of the dandy who makes of his body, his behavior, his feelings and passions, his very existence, a work of art. Modern man, for Baudelaire, is not the man who goes off to discover himself, his secrets and his hidden truth; he is the man who tries to invent himself. This modernity does not "liberate man in his own being"; it compels him to face the task of producing himself.

4. Let me add just one final word. This ironic heroization of the present, this transfiguring play of freedom with reality, this ascetic elaboration of the self—Baudelaire does not imagine that these have any place in society itself, or in the body politic. They can only be produced in another, a different place, which Baudelaire calls art.

I do not pretend to be summarizing in these few lines either the complex historical event that was the Enlightenment, at the end of the eighteenth century, or the attitude of modernity in the various guises it may have taken on during the last two centuries.

I have been seeking, on the one hand, to emphasize the extent to which a type of philosophical interrogation—one that simultaneously problematizes man's relation to the present, man's historical mode of being, and the constitution of the self as an autonomous subject—is rooted in the Enlightenment. On the other hand, I have been seeking to stress that the thread that may connect us with the Enlightenment is not faithfulness to doctrinal elements, but rather the permanent reactivation of an attitude—that is, of a philosophical ethos that could be described as a permanent critique of our historical era. I should like to characterize this ethos very briefly.

A. NEGATIVELY

1. This ethos implies, first, the refusal of what I like to call the "blackmail" of the Enlightenment. I think that the Enlightenment, as a set of political,

economic, social, institutional, and cultural events on which we still depend in large part, constitutes a privileged domain for analysis. I also think that as an enterprise for linking the progress of truth and the history of liberty in a bond of direct relation, it formulated a philosophical question that remains for us to consider. I think, finally, as I have tried to show with reference to Kant's text, that it defined a certain manner of philosophizing.

But that does not mean that one has to be "for" or "against" the Enlightenment. It even means precisely that one has to refuse everything that might present itself in the form of a simplistic and authoritarian alternative: you either accept the Enlightenment and remain within the tradition of its rationalism (this is considered a positive term by some and used by others, on the contrary, as a reproach); or else you criticize the Enlightenment and then try to escape from its principles of rationality (which may be seen once again as good or bad). And we do not break free of this blackmail by introducing "dialectical" nuances while seeking to determine what good and bad elements there may have been in the Enlightenment.

We must try to proceed with the analysis of ourselves as beings who are historically determined, to a certain extent, by the Enlightenment. Such an analysis implies a series of historical inquiries that are as precise as possible; and these inquiries will not be oriented retrospectively toward the "essential kernel of rationality" that can be found in the Enlightenment and that would have to be preserved in any event; they will be oriented toward the "contemporary limits of the necessary," that is, toward what is not or is no longer indispensable for the constitution of ourselves as autonomous subjects.

2. This permanent critique of ourselves has to avoid the always too facile confusions between humanism and Enlightenment.

We must never forget that the Enlightenment is an event, or a set of events and complex historical processes, that is located at a certain point in the development of European societies. As such, it includes elements of social transformation, types of political institution, forms of knowledge, projects of rationalization of knowledge and practices, technological mutations that are very difficult to sum up in a word, even if many of these phenomena remain important today. The one I have pointed out and that seems to me to have been at the basis of an entire form of philosophical reflection concerns only the mode of reflective relation to the present.

Humanism is something entirely different. It is a theme or rather a set of themes that have reappeared on several occasions over time in European societies; these themes, always tied to value judgments, have obviously varied greatly in their content, as well as in the values they have preserved. Furthermore, they have served as a critical principle of differentiation. In the seventeenth century, there was a humanism that presented itself as a critique of Christianity or of religion in general; there was a Christian humanism opposed to an ascetic and much more theocentric humanism. In the nineteenth century, there was a suspicious humanism, hostile and critical toward science, and another that, to the contrary, placed its hope in that same science. Marxism has been a humanism; so have existentialism and personalism; there was a time when people supported the humanistic values represented by National Socialism, and when the Stalinists themselves said they were humanists.

From this, we must not conclude that everything that has ever been linked with humanism is to be rejected, but that the humanistic thematic is in itself too supple, too diverse, too inconsistent to serve as an axis for reflection. And it is a fact that, at least since the seventeenth century, what is called humanism has always been obliged to lean on certain conceptions of man borrowed from religion, science, or politics. Humanism serves to color and to justify the conceptions of man to which it is, after all, obliged to take recourse.

Now, in this connection, I believe that this thematic, which so often recurs and which always depends on humanism, can be opposed by the principle of a critique and a permanent creation of ourselves in our autonomy: that is, a principle that is at the heart of the historical consciousness that the Enlightenment has of itself. From this standpoint I am inclined to see Enlightenment and humanism in a state of tension rather than identity.

In any case, it seems to me dangerous to confuse them; and further, it seems historically inaccurate. If the question of man, of the human species, of the humanist, was important throughout the eighteenth century this is very rarely I believe because the Enlightenment considered itself a humanism. It is worthwhile, too, to note that throughout the nineteenth century the historiography of sixteenth-century humanism, which was so important for people like Saint-Beuve or Burckhardt, was always distinct from and sometimes explicitly opposed to the Enlightenment and the eighteenth century. The nineteenth century had a tendency to oppose the two at least as much as to confuse them.

In any case, I think that just as we must free ourselves from the intellectual blackmail of being "for or against the Enlightenment," we must escape from the historical and moral confusionism that mixes the theme of humanism with the question of the Enlightenment. An analysis of their complex relations in the course of the last two centuries would be a worthwhile project, an important one if we are to bring some measure of clarity to the consciousness that we have of ourselves and of our past.

B. POSITIVELY

Yet while taking these precautions into account, we must obviously give a more positive content to what may be a philosophical ethos consisting in a critique of what we are saying, thinking, and doing, through a historical ontology of ourselves.

 1. This philosophical ethos may be characterized as a *limit-attitude*. We are not talking about a gesture of rejection. We have to move beyond the outside-inside alternative; we have to be at the frontiers. Criticism indeed consists of analyzing and reflecting upon limits. But if the Kantian question was that of knowing what limits knowledge has to renounce transgressing, it seems to me that the critical question today has to be turned back into a positive one: in what is given to us as universal, necessary, obligatory, what place is occupied by whatever is singular, contingent, and the product of arbitrary constraints? The point, in brief, is to transform the critique conducted in the form of necessary limitation into a practical critique that takes the form of a possible transgression.

This entails an obvious consequence: that criticism is no longer going to be practiced in the search for formal structures with universal value, but rather as a historical investigation into the events that have led us to constitute ourselves and to recognize ourselves as subjects of what we are doing, thinking, saying. In that sense, this criticism is not transcendental, and its goal is not that of making a metaphysics possible: it is genealogical in its design and archaeological in its method. Archaeological—and not transcendental—in the sense that it will not seek to identify the universal structures of all knowledge or of all possible moral action, but will seek to treat the instances of discourse that articulate what we think, say, and do as so many historical events. And this critique will be genealogical in the sense that it will not deduce from the form of what we are what it is impossible for us to do and to know; but it will separate out, from the contingency that has made us what we are, the possibility of no longer being, doing, or thinking what we are, do, or think. It is not seeking to make possible a metaphysics that has finally become a science; it is seeking to give new impetus, as far and wide as possible, to the undefined work of freedom.

2. But if we are not to settle for the affirmation or the empty dream of freedom, it seems to me that this historico-critical attitude must also be an experimental one. I mean that this work done at the limits of ourselves must, on the one hand, open up a realm of historical inquiry and, on the other, put itself to the test of reality, of contemporary reality, both to grasp the points where change is possible and desirable, and to determine the precise form this change should take. This means that the historical ontology of ourselves must turn away from all projects that claim to be global or radical. In fact we know from experience that the claim to escape from the system of contemporary reality so as to produce the overall programs of another society, of another way of thinking, another culture, another vision of the world, has led only to the return of the most dangerous traditions. I prefer the very specific transformations that have proved to be possible in the last twenty years in a certain number of areas that concern our ways of being and thinking, relations to authority, relations between the sexes, the way in which we perceive insanity or illness; I prefer even these partial transformations that have been made in the correlation of historical analysis and the practical attitude, to the programs for a new man that the worst political systems have repeated throughout the twentieth century.

I shall thus characterize the philosophical ethos appropriate to the critical ontology of ourselves as a historico-practical test of the limits that we may go beyond, and thus as work carried out by ourselves upon ourselves as free beings.

3. Still, the following objection would no doubt be entirely legitimate: if we limit ourselves to this type of always partial and local inquiry or test, do we not run the risk of letting ourselves be determined by more general structures of which we may well not be conscious, and over which we may have no control?

To this, two responses. It is true that we have to give up hope of ever acceding to a point of view that could give us access to any complete and definitive knowledge of what may constitute our historical limits. And from this point of view the theoretical and practical experience that we have of our limits and of the possibility of moving beyond them is always limited and determined; thus we are always in the position of beginning again.

But that does not mean that no work can be done except in disorder and contingency. The work in question has its generality, its systematicity, its homogeneity, and its stakes.

(a) Its Stakes

These are indicated by what might be called "the paradox of the relations of capacity and power." We know that the great promise or the great hope of the eighteenth century, or a part of the eighteenth century, lay in the simultaneous and proportional growth of individuals with respect to one another. And, moreover, we can see that throughout the entire history of Western societies (it is perhaps here that the root of their singular historical destiny is located—such a peculiar destiny, so different from the others in its trajectory and so universalizing, so dominant with respect to the others), the acquisition of capabilities and the struggle for freedom have constituted permanent elements. Now the relations between the growth of capabilities and the growth of autonomy are not as simple as the eighteenth century may have believed. And we have been able to see what forms of power relation were conveyed by various technologies (whether we are speaking of productions with economic aims, or institutions whose goal is social regulation, or of techniques of communication): disciplines, both collective and individual, procedures of normalization exercised in the name of the power of the state, demands of society or of population zones, are examples. What is at stake, then, is this: How can the growth of capabilities be disconnected from the intensification of power relations?

(b) Homogeneity

This leads to the study of what could be called "practical systems." Here we are taking as a homogeneous domain of reference not the representations that men give of themselves, not the conditions that determine them without their knowledge, but rather what they do and the way they do it. That is, the forms of rationality that organize their ways of doing things (this might be called the technological aspect) and the freedom with which they act within these practical systems, reacting to what others do, modifying the rules of the game, up to a certain point (this might be called the strategic side of these practices). The homogeneity of these historico-critical analyses is thus ensured by this realm of practices, with their technological side and their strategic side.

(c) Systematicity

These practical systems stem from three broad areas: relations of control over things, relations of action upon others, relations with oneself. This does not mean that each of these three areas is completely foreign to the others. It is well known that control over things is mediated by relations with others; and relations with others in turn always entail relations with oneself, and vice versa. But we have three axes whose specificity and whose interconnections have to be analyzed: the axis of knowledge, the axis of power, the axis of ethics. In other terms, the historical ontology of ourselves has to answer an open series of questions; it has to make an indefinite number of inquiries which may be multiplied and specified as much as we like, but which will all address the questions systematized as follows: How are we constituted as subjects of our own knowledge? How are we constituted as subjects who exercise or submit to power relations? How are we constituted as moral subjects of our own actions?

(d) Generality

Finally, these historico-critical investigations are quite specific in the sense that they always bear upon a material, an epoch, a body of determined practices and discourses. And yet, at least at the level of the Western societies from which we derive, they have their generality, in the sense that they have continued to recur up to our time: for example, the problem of the relationship between sanity and insanity, or sickness and health, or crime and the law; the problem of the role of sexual relations; and so on.

But by evoking this generality, I do not mean to suggest that it has to be retraced in its metahistorical continuity over time, nor that its variations have to be pursued. What must be grasped is the extent to which what we know of it, the forms of power that are exercised in it, and the experience that we have in it of ourselves constitute nothing but determined historical figures, through a certain form of problematization that defines objects, rules of action, modes of relation to oneself. The study of [modes of] problematization (that is, of what is neither an anthropological constant nor a chronological variation) is thus the way to analyze questions of general import in their historically unique form.

A brief summary, to conclude and to come back to Kant.

I do not know whether we will ever reach mature adulthood. Many things in our experience convince us that the historical event of the Enlightenment did not make us mature adults, and we have not reached that stage yet. However, it seems to me that a meaning can be attributed to that critical interrogation on the present and on ourselves which Kant formulated by reflecting on the Enlightenment. It seems to me that Kant's reflection is even a way of philosophizing that has not been without its importance or effectiveness during the last two centuries. The critical ontology of ourselves has to be considered not, certainly, as a theory, a doctrine, nor even as a permanent body of knowledge that is accumulating; it has to be conceived as an attitude, an ethos, a philosophical life in which the critique of what we are is at one and the same time the historical analysis of the limits that are imposed on us and an experiment with the possibility of going beyond them.

This philosophical attitude has to be translated into the labor of diverse inquiries. These inquiries have their methodological coherence in the at once archaeological and genealogical study of practices envisaged simultaneously as a technological type of rationality and as strategic games of liberties; they have their theoretical coherence in the definition of the historically unique forms in which the generalities of our relations to things, to others, to ourselves, have been problematized. They have their practical coherence in the care brought to the process of putting historico-critical reflection to the test of concrete practices. I do not know whether it must be said today that the critical task still entails faith in Enlightenment; I continue to think that this task requires work on our limits, that is, a patient labor giving form to our impatience for liberty.

Historical Problematizations

The "Trial" of Theoretical Curiosity

Hans Blumenberg

Hans Blumenberg (1920–1996) was a German philosopher and historian of ideas. He was particularly concerned with the transition and transformation in philosophy, science, and theology from medieval Europe to the modern age. Blumenberg offers an erudite analysis of the ethos of modernity after the seventeenth century and insists on the originality and legitimacy of this ethos. The "'trial' of theoretical curiosity," described in the essay included here, focuses on the problem of what he calls modern scientific self-assurance as well as the anthropological limits and blind spots of scientific enlightenment.

According to a simple formula for mirroring backgrounds, the statement that the contemporary world can exist only *by means of* science stands in suspicious relation to the fact that this is asserted by people who themselves make their living *from* science. But this suspicion is still harmless compared to the suspicion that results from the fact that science itself brought forth the very world, to live in which depends on—and makes us increasingly dependent on—science's continued existence and continued operation. The dilemma of any attempt to focus on this underlying state of affairs lies in the fact that talk *about* science only begets a further science [*Wissenschaft*: knowledge], whatever one chooses to call it. Nor can the attempts to inquire back into a prescientific sphere, whether synchronically, in the "lifeworld," or diachronically, in history, free themselves from this adhesion. The great gesture of self-liberation is no help here. If one wants to speak of theoretical curiosity as one of the motivating forces of the process of science, then one cannot escape entanglement in the misgiving that one is being swept along oneself in the stream of that motivating force. It is curiosity that draws one's attention to curiosity; curiosity depends entirely on itself to throw off the discrimination imposed on it, as its modern history shows. It is not able to confirm the Platonic hope that one could know in advance what is held in store for it.

That the difficulties we have, and will increasingly have, with science are always integrated into it as scientific difficulties is only one aspect of the outlook on the subject of "theoretical curiosity": the inevitability of a failure to find an Archimedean point over against the reality of science. The other aspect is that of responsibility. Most of the people whose lives today depend on science would not even be alive, or would no longer be alive, if science had not made their lives possible and prolonged them. When one puts it that way, it sounds laudable. On the other hand, this means at the same time that the overpopulation of our world is also an excess produced by science. Are there unambiguous conclusions that can be drawn from this statement? One should avoid too easy answers to this question. To a large extent, science has broken the brutal mechanism of the "survival of the fittest": it gives more life to people who are less "fit" for life and keeps them alive longer. Is this a humane achievement? Here again it would be frivolous to say that we have an answer to the question. But to pose it is to make as clear as possible the significance of what one is dealing with when one not only focuses on the dependence of our reality on science but also defines that dependence as problematic. The fact that, biologically speaking, we no longer live in a Darwinian world, or at any rate we live in a world that is less and less Darwinian, is a consequence of science that, even if in its turn it has consequences that are not evident at a glance, is simply irreversible. Science integrates into itself even the responsibility for the consequences of its consequences, by itself giving the alarm.

The existence, and even the mere dimensions of the existence, of science are not things over which we have the power of disposition as long as we do not feel entitled to answer in the affirmative the question whether the non-existence of existing persons or the discontinued existence of people whose existence has at any time been in danger would have been a more humane alternative. The only person who can presume to play with the idea of a discontinuance or a reduction in the human effort called "science" is one who has a low estimate of the susceptibility of the motivation of theory to disturbance. The limits of responsible behavior may be much narrower here than many people imagine. Between uneasiness about science's autonomous industry and the constraints resulting from its indispensability lies an indeterminate latitude within which we are free to act as we wish but that it would be misleading to project upon science as a whole.

The difficulties that we have with science and the rule of those who represent science [*die Epistemokratie*] suggest the gleam of a hope that we might escape them by setting up yet another science, an "ultimate science," which would concern itself with nothing but science itself. Another thing that makes this idea attractive is that it promises arbitrative functions, the exercise of power over the powerful, even if only over people defined as such for the purposes of the arbitration alone. It would be the Archimedean point—or else the exponential increase, through their iteration, of all the difficulties we already have with science. Why, after all, should a "science of science," which elevates itself to the job of the emphatically so-called critique of every other species of science, be free from the problematic that it would be sure to find in them? The discernment of a need for such a metadiscipline, the consensus regarding its acute urgency, imply nothing whatsoever about its possibility. But skepticism becomes all the more irritating, the nearer we seem to be to filling the office of arbiter.

We cannot live without science. But that is itself largely an effect produced by science. It has made itself indispensable. But what this observation does not explain is what it was that set the "industry" of science going and keeps it in motion. On the contrary, there exists a peculiar uncertainty as to what the motives are that move and intensify this epochal effort. One extreme is the mechanical connection between autonomous industriousness and meaninglessness that Victor Hugo expressed in 1864: "Science searches for perpetual motion. It has found it; it is itself."[1] The absolute necessity of science in the contemporary world does not license any inferences about the process by which it began. Even if existential exigencies prevent us from interrupting the functioning of science, this is not enough to show that its reality originates in its necessity for life. We must reckon with a break in motivation when the moving impetus of theory no longer comes directly from the "lifeworld"—from the human interest in orientation in the world, the will to the expansion of effective reality, or the need for the integration of the unknown into the system of the known. This is where uneasiness sets in. Necessity is manifestly not enough; it is unable to dispel the suspicion of meaninglessness or, perhaps, even more severely, the "fear of a total meaninglessness that lies behind every science."[2]

The talk of the "science industry" [*Wissenschaftsbetrieb*] that has become popular refers, of course, to the objective structural similarity of scientific institutions and processes to those of industry, but in its most extreme form it points above all, contemptuously, to the bustling and autonomous industry of scientific work as it is now organized, to the rupture of the connection between a motivation for the theoretical attitude that is founded in the "lifeworld" and the realization of that attitude under the conditions on which the effectiveness of modern science depends; and finally it also points to the lack of congruence between the outputs of the autonomous process and the expectations, rooted in the European tradition, that the truth would make men happy and free. Seen from the point of view of the conception of theory that corresponds to these expectations, the connection between science and securing the chance to live is really an unexpected historical development. This surprise is not the sole cause of our uneasiness with science, but it is an essential element in the situation.

Since ancient times, what theory was supposed to do was not to make life possible but to make it happy. Hence also the first epochal injection of mistrust in theory, when happiness had become a matter for hope directed at the next world, for a salvation that man could not bring about, though it was still defined as *visio beatifica* [beatific vision]—as the acquisition of truth, fulfillment through theory. The premise that only the final possession of truth could guarantee man's happiness went over from ancient thought into the interpretation of the biblical eschatology. That life was pleasanter for one who knew than for one who sought knowledge was a premise Aristotle took for granted; it corresponded to his concept of God and especially to his physics of finite space and thus of finite "natural" motions justified only by—and ending in—a goal state of rest. The early-modern renewal of the pretension to unrestricted theoretical curiosity turned against the exclusion of pure theory, and of the pure happiness that was bound up with it, from the realm of what could be reached in this world, just as it turned against the medieval God's claim to exclusive insight into nature as His work. The investigator of nature could no longer remain—nor again become—the ancient world onlooker, though he had to reconstruct the connection between cognitive truth and finding happiness in a different way if, following Francis Bacon's new formula, domination over nature was to be a precondition of the recovery of paradise.

From a central affect of consciousness there arises in the modern age an indissoluble connecting link between man's historical self-understanding and the realization of scientific knowledge as the confirmation of the claim to unrestricted theoretical curiosity. The "theoretical attitude" may be a constant in European history since the awakening of the Ionians' interest in nature; but this attitude could take on the explicitness of insistence on the will and the right to intellectual curiosity only after it had been confronted with opposition and had had to compete with other norms of attitude and fulfillment in life. Just as "purity" as a quality of the theoretical attitude could only be formulated in the circumstances of Plato's opposition to the Sophists' instrumentalization of theory, so also the "right" to an unrestricted cognitive drive constituted itself and was united with the self-consciousness of an epoch only after the Middle Ages had discriminated against such intellectual pretensions and put them in a restrictive adjunct relation to another human existential interest posited as absolute. The rehabilitation of theoretical curiosity at the beginning of the modern age is just not the mere renaissance of a life ideal that had already been present once before and whose devaluation, through the interruption of its general acceptance, had only to be reversed.

The classical anthropological question whether man strives for knowledge on account of an inner and uncoerced impulse of his nature or whether the necessity of gaining knowledge is thrust upon him by the naked demands of the prolongation of his existence is no timeless, unhistorical problematic, although its continual recurrence—for instance, in the contrast between Husserl's phenomenological radical of the "theoretical attitude" and Heidegger's Daseins-analytic "existentiale" of "care"—seems to make this a natural assumption. The most widely read handbook of physics in the century of the Enlightenment was able to harmonize this question with no trouble: "Necessity and men's curiosity have perhaps made equal contributions to the discovery and further elaboration of the science of nature. . . ."[3] Jürgen Mittelstrass has proposed a distinction between "naive" and "reflected" curiosity,[4] at the same time describing "talk of a novel type of curiosity that initiates the modern age itself"—of a "self-conscious curiosity"—as "unsatisfactory" "so long as this beginning of the modern age" cannot be distinguished by "specific transactions."[5] I am not going to go into the question whether one who demands evidence of "specific transactions" is not left in the hands of an overdetermined concept of history and thus in continued bondage to the criteria of official documentation [*Aktenkundigkeit*]. The proposed distinction in any case seems to me to be useful.

Just as anyone who wants to characterize the modern age as an epoch marked by technology or tending toward that end finds his attention directed again and again to technicity as an original anthropological characteristic and thus as an omnipresent human structure, which admits only a quantitative differentiation of increased complexity between a stone tool and a moon rocket—so the stress on the element of curiosity undergoes the same process: Curiosity is a mark of youthfulness even in animals, and a mark, all the more, of man as the animal who remains youthful. Naive curiosity, then, would be the constant; but at the same time it is the substratum around which historical articulation and focus set in.

It is just this process that is my subject here: As a result of the discrimination against it, what was natural and went without saying is explicitly "entered into" and accentuated; play with the world's immediacy becomes the seriousness of methodical formation, the necessity [*Notwendigkeit*] of self-preservation becomes the versatility [*Wendigkeit*] of self-assertion, and what was a mere occupation becomes a prerogative to be secured and at the same time becomes the energy that increases exponentially each time it turns out that the suspected reservation of the unknown but knowable does exist— that knowledge can extend beyond the Pillars of Hercules, beyond the limits of normal optics and the postulate of visibility, in other words, beyond the horizons that had been assigned to man as long as he had thought that he could remain the onlooker in repose, the leisurely enjoyer of the world, taken care of by providence. The interpretation of natural restrictions as representing a realm to which man was denied access "in this world" radically altered the quality of the theoretical form of life recommended by ancient philosophy.

To demonstrate the logic of this process is immediately to exclude the naturalistic suggestion that in the preponderance of theoretical curiosity in the modern age what we confront is a fateful recurrence of the same, the turning of an anthropological tide. Toward the end of the nineteenth century Otto Liebmann exhibited the satisfaction of one who had finally pinned down a law of nature in explaining the epochs of theory:

> That is, the propensity to theorizing seems to be subject, like other human inclinations, to an alternation of ebb and flood. In the causal context of a great variety of cultural-historical factors it experiences its alternating maxima and minima. There are ages in which it swells into a regular monomania and overruns in hypertrophic fashion the more modest need for the gathering of simple observational knowledge of matters of fact. There are other ages in which it sinks below the zero point and seems entirely overcome by that same (to it) antagonistic need. When a doctrinaire attitude, ensconced in what has become a rigid and dogmatically closed world view, considers itself to have arrived at the summit of wisdom and now employs all its sagacity in elaborating all the subtlest ramifications of the finished conceptual system that it holds to be true—but equally, however, when a period that is carried away with youthful hope, a reforming period, in its precipative drive to give form, peoples the unknown land of anticipated truth with hitherto unimagined mental creations and strays into the boundless and the fantastic—then the reaction against such hyper-theorizing follows in natural sequence, and subsequent generations, cautiously assessing the evidence, will have to invest half of their effort in the critical clearing out of overflowing Augean stalls. Then, to be sure, that excess is followed by a deficiency; a praiseworthy avoidance of doctrinaire illusions, an understandable fear of unreliable pseudo-theories, an entirely admirable feeling for the truth causes people to fall into the other extreme. . . .[6]

This naturalistic approach makes very clear, in negative form, what should be expected of a presentation of the historical "proceedings" relative to curiosity that aims at rational analysis.

Our situation is not that of the beginning of the modern age, however distinguishable by "specific transactions" that beginning may be. Is the problem of making a beginning still our problem? Jürgen Mittelstrass has answered this question by giving his concept of "reflected curiosity" a specifically heterogeneous function that I would like to characterize as that of an already iterated "reflection": What set the modern age's curiosity in motion no longer needs—in its self-accelerated, immanently propelled motion—rehabilitation and restitution of its primary energy; it has become indifferent to the new, as such, on account of its experience of the latter's inevitability which may even constitute for it a burden to be endured, and instead it is all the more sensitive to the direction that belongs to the motion that is thus stimulated, sensitive to the question of where it is headed.

In this situation, anyone at all who "defends man's interest in what, so to speak, does not concern him" seems anachronistic—unless perhaps this were once again an act of defending theoretical curiosity in circumstances where it was supposed to interest itself only in material that did not run counter to powerful interests. Even in the categorization of theory as a derivative attitude subordinate to the radical of "care," there is a possibility, if not a necessity, of requiring the interest in theory to legitimate itself once again by demonstrating a contemporary and relevant, or even an authoritatively prescribed, "care" as its source. Scarcely a decade after theory, as mere gaping at what is "present at hand," had been, if not yet despised, still portrayed as a stale recapitulation of the content of living involvements, it was the greatness of the solitary, aged Edmund Husserl, academically exiled and silenced, that he held fast to the resolution to engage in theory as the initial act of European humanity and as a corrective for its most terrible deviation, and that he required of it a rigorous consistency, which is still, or once again, felt to be objectionable. Hermann Lübbe has described as the characteristic mark of this philosophizing, especially in the late works, the "rationalism of theory's interest in what is without interest":

> The existential problem of a scholar who in his old age was forbidden to set foot in the place where he carried on his research and teaching never shows through, and even the back of the official notice that informed him of this prohibition was covered by Husserl with philosophical notes. That is a case of "carrying on" whose dignity equals that of the sentence, "*Noli turbare circulos meos*" [Don't disturb my circles].[7,8]

The bearing of the concept of "reflected" curiosity is on neither the propulsion of theoretical activity nor its resistance to commands that it halt or demonstrate its relevance; that is, the bearing is not on world orientation but on the orientation of the total process of the curiosity that is consciously formed out of its earlier naiveté. Its expectation no longer relates to "the discovery of something entirely new but rather to the now never ending question, what will come next."[9] The self-conscious curiosity that, at the beginning of the modern age, at first turned against history as the epitome of the abrogation of reason and of preoccupation with prejudices and impenetrable reserves made its own history—as soon as it had one—a subject of inquiry, not by adopting a qualitatively new attitude but because it still possessed the naive ubiquitousness that looks under every stone and over every fence, and consequently also into its own records. Reflection [*Reflexion*] first arises as a result of the resistance that an examination of the history of science as a supposedly linear process of accomplishment opposes to the naive assumption that it is an "object" like any other. Reflection on where one actually finds oneself and on what should come next is a side effect of the "encyclopedic" impulse and activity that are aimed at taking stock, at still keeping control of what can no longer be surveyed and taken in all at once, putting it in usable form as an available potential. Curiosity acquires its conservative complement in the encyclopedic program: It cannot understand itself only as motion; it must also seek to grasp its topography, the boundaries that are no longer set for it by an external authority but that instead it itself describes by the totality of its findings. Diderot's article "Encyclopédie," written in 1755, marks the onset of reflection on the competition between the need to survey and assimilate—to take stock—and the need to orient further progress. For the organizer of the *Encyclopédie* the problematic of the use of time is clear: It is true that assimilation does not yet take longer than the duration of the validity of what is assimilated, but the fundamental encyclopedic ideas of universal accessibility and of replicability through organization become, at least, questionable. On the one hand, Diderot is confident that future generations will be able to construct a better encyclopedia on the basis of his; but on the other hand, he emphasizes the special circumstances that have made this particular work possible, and thus the uncertainty whether comparable conditions will be present in the future. The balance of these considerations reads as follows: "L'Encyclopédie peut aisément s'améliorer; elle peut aussi aisément se détériorer." [The Encyclopedia can easily be improved; it can just as easily deteriorate.]

The initial success of theoretical curiosity in the modern age would have been inconceivable without the transition from "naive" to "self-conscious" curiosity. The latter had not only emerged through its competition with the concern for salvation and its argument with the transcendent reservation [of realms of knowledge]; once people had presumed to peek behind the back cloth ["behind the scenes"] of creation, it had also been able to translate the results, as confirmations of its suspicions as well as of its right to what was withheld, into the energy of the *Plus ultra* [Still further: Francis Bacon]. This dynamic of self-confirmation freed curiosity from the connotations of a "base instinct" that bound man's attention to inessential and superficial matters, to prodigies, monstrosities—in fact, to *curiosa* [curiosities]. But the very summing up of these confirming effects engendered a need that one could classify, initially, as "topographical."

The dilemma of the idea of the encyclopedia makes it clear why "re-flected" curiosity will find the dynamic set in motion by the self-conscious interest in knowledge objectionable: The expansion of the horizon of known and understood reality could not be coordinated with the presence of what was already accessible within this horizon. Diderot did indeed think of the perfection of the encyclopedia as an objective task for the future; he thought about what abilities the contributors would have to possess, what circum-stances would be favorable; but he did not consider the situation of the reader and the way in which it would be altered by the universal quality of the work. He would already have been able to say what we have to say today, that while we know more about the world than we ever did before, this "we" does not by any means mean "I." The "we" of this statement confronts the "I" only in the form of institutions—of encyclopedias, academies, universities. These represent higher-level agencies [*Übersubjekte*] that administer knowledge about reality in space and time and organize its growth. The disproportion between what has been achieved in the way of theoretical insight into reality and what can be transmitted to the individual for his use in orienting himself in his world is disconcertingly unpreventable. But the intensity of the pro-cess becomes critical in regard to not only the relation between the objective stock of knowledge and its translatability into subjective orientation but also the stability of that stock itself in view of the fact that in the succession of generations of knowledge, the length of the "half-life" of each, on its way to obsolescence, has already dropped to less than a decade. The phrase "in possession of the truth" [*Wahrheitsbesitz*]—no matter how one defines truth epistemologically—is no longer capable of nonironical employment. Even if, forgoing the use of the classical term, one speaks of the encyclopedic postulate of possessing the greatest possible stock of information, still the accelerating rate at which that information decays means that the individual is compelled to acquire a capacity for provisional relations with it, for transitory reliance on it, within the duration of his individual lifetime. It is easy to imagine this disappointment with the stability of scientific knowledge pushing people toward modes of "having" theoretical propositions that seem less unstable and less taxing because they are hardly falsifiable.

This phenomenon of the acceleration of the theoretical process can no longer be explained by reference to the hyperfunctioning of a theoretical curiosity that organized itself around the recovery of the right to unrestricted expansion. Undoubtedly there exists somewhere in the course of the progressive consolidation of this structured process a point at which the possibility of the intervention of exogenous, "lifeworldly," historical motivations narrows and finally disappears and after which one has to say that in relation to what comes about as science, and to what scientists do, theoretical curiosity is now only a secondary factor. However much it may still determine the genesis of a choice of profession, it has correspondingly little effect on the objective state and the availability to the individual of the structured process in reality. This also—the lack of room for individual motivation, for an authentic initiative—is involved in our uneasiness with regard to science.

Of course it will not be possible to determine the exact point in time up to which, while an individual might not have been able to assimilate the totality of the truths accessible at the time—that limiting case has probably always been unattainable—still, enough could be attained in one lifetime that the individual could impute to himself a substantial share in what was known of reality and what seemed in any way necessary for its comprehension. It is only on this assumption, that the truth in its totality was at the disposition of the individual, that the ancient association of eudaimonia with theory, as its precondition, could be held on to and even renewed. For when the program of a science safeguarded by method was projected in the early modern age, this elementary assumption was renewed. The reality in which man, both as individual and as society, lived was supposed to remain identical with the reality that theoretical knowledge was to open up and make perspicuous for human action. Otherwise Descartes could not have promised the definitive morality as the consequence of the perfection of physics.

The definitive morality, which as the epitome of materially adequate behavior was supposed to guarantee human happiness, remained bound to the continuing presence of the perfected theory for practice because the behavioral norm in each case emerged from personal insight into the structure of reality. But almost simultaneously, in Francis Bacon, a concept of human happiness appeared that separated theory from existential fulfillment by reducing the necessary knowledge to the amount fixed by the requirements of domination over natural reality. The recovery of paradise was not supposed to yield a transparent and familiar reality but only a tamed and obedient one. For this equivalent of a magic attitude to reality, the individual no longer needed to understand himself in his relation to reality; instead it was sufficient if the combination of everyone's theoretical accomplishments guaranteed a state of stable domination over this reality, a state of which the individual could be a beneficiary even without having insight into the totality of its conditions. The subject of theory and the subject of the successful life no longer needed to be identical. This appears as relief from a demand that was immediately to become unfulfillable, even before the incongruence between theoretical objectivity and individual competence had become foreseeable.

Here it has already become in principle possible and permissible for scientific knowledge to be an instrument of specialists, a reserve administered by initiates, institutionalized not as something one can possess but rather as an available potential. Theoretical curiosity serves only to guarantee that in spite of the impediments to it, the cognitive process gets under way and is pressed forward; but the vindication of its self-assertion is not accomplished by the mere fact that the overstepping of the boundaries of the known world, which it dares for the first time, does enable it to discover new worlds but only by a demonstration of the real usefulness of knowledge as a source of capability. This is the explanation for the delay that intervened before Bacon's theory of theory enjoyed real success. His ulterior magical conception—that a world that was created by the word must also be one that can be dominated by the word, that to be in paradise means to know the names of things—had to be forgotten. This is not the context in which, for instance, Montesquieu speaks of the curiosity that is inherent in all men in his address to the Académie (near the beginning of the eighteenth century) on the motives to encourage people to pursue the sciences: that curiosity, he says, has never been so well vindicated as in the present age, in which one daily hears it said that the limits of man's knowledge are being infinitely expanded and that the savants themselves are so amazed at what they know that sometimes they doubt the reality of their successes.[10]

To understand the process of the legitimation of theoretical curiosity as a basic feature of the history of the beginning of the modern age certainly does not mean to make curiosity into the "destiny" of history, or one of its absolute values. The legitimacy of the modern age is not the legitimation of its specific constituent elements under all possible circumstances. It is possible that Socrates was in the right when, as Cicero says, for the first time he brought philosophy down from the heavens, settled it in the cities, introduced it into people's homes, and forced it to investigate life, manners, and norms of behavior.[11] But one must also see what this Socratic turning became once it ceased to be understood as making man the subject of inquiry and was interpreted instead as the theological reservation of other subjects to divine sovereignty and was accordingly placed in Socrates' mouth as the abbreviated motto, *Quae supra nos, nihil ad nos* [What is above us is nothing to us].[12] The rehabilitation of theoretical curiosity is justified in the first instance only as the rejection of discrimination against it, of a restriction of its legitimacy that had only initially been grounded on concern for the salvation of the one who was thus constrained. Despite the fact that the connection between theory and eudaimonia that was established in antiquity was involved in the process of the emancipation of curiosity at the beginning of the modern age, the question whether man in fact achieved happiness too in exercising the rights that he had thus recovered has no bearing on the establishment of the legitimacy of his pretension: From the perspective of a pure eudaimonism, restrictions on human inclinations, based on any principle whatever, are incompatible with the motivation that is proper to the desire for happiness; that fact defines the burden of proof that has to be met in justifying restrictive reservations. The Socratic question whether man's interest in himself does not require neglect of his interest in nature does provide a form of argument for the discrimination against curiosity but is not characteristic of the state of affairs at the close of the Middle Ages, which was the determining factor in the formation of the new pretension. The balance sheet of theoretical curiosity in general is not predetermined by its legitimacy in the modern age. Still, the latter does provide food for thought that is relevant to the former, as is proper for a piece of philosophical reflection.

Justifications of Curiosity as Preparation for the Enlightenment

Hans Blumenberg

Hans Blumenberg (1920–1996) was a German philosopher and historian of ideas. He was particularly concerned with the transition and transformation in philosophy, science, and theology from medieval Europe to the modern age. Continuing his erudite diagnosis of the originality and legitimacy of new modes of thinking and inquiry after the seventeenth century, in this essay Blumenberg accounts for the distinctively modern arguments justifying the right to intellectual curiosity and the free pursuit of scientific knowledge.

The modern age has understood itself as the age in which reason, and thus man's natural vocation, definitively prevailed. The difficulty created by this self-interpretation was to explain the delayed appearance in history of the form of existence that, as a result of its identity with the nature of man, should have been ubiquitous and taken for granted throughout history. The conceptions of the historical process that try to overcome this problem can be categorized according to whether they describe the prehistory of the age of reason as the natural impotence or the forcible suppression of the power of rationality. Like many metaphors, those of the organic growth of rationality or the coming of the light of day after the long night of its absence have an initial but not a lasting plausibility. The idea of a continuous upward progress of rationality contradicted the fundamental idea of the radical, revolutionary self-empowerment of reason as an event of epoch-making, unexpected suddenness. The idea of reason liberating itself from its medieval servitude made it impossible to understand how such a servitude could ever have been inflicted upon the constitutive power of the human spirit and could have continued in force for centuries. Another dangerous implication of this explanation was that it was bound to inject doubt into the self-consciousness of reason's definitive victory and the impossibility of a repetition of its subjugation. Thus the picture of its own origin and possibility in history that the epoch of rationality made for itself remained peculiarly irrational. In the Enlightenment's understanding of history, the relation between the Middle Ages and the modern age is characteristically dualistic, and this is expressed more than anywhere else in the conception formulated by Descartes of an absolute, radical new beginning, whose only prerequisites lay in the rational subject's making sure of itself, and for which history could become a unity only under the dominance of "method."

One could say that historical understanding and the historical attitude were formed precisely to the extent that this dualism, presupposed by the modern age's pretension to spontaneity, was overcome and the Middle Ages were brought into a unified conception of history. But this result was achieved through the effacement of the epochal threshold by the demonstration that elements of what was supposedly new could be traced back to factors that could be shown to be "already" present in the makeup of the Middle Ages. The transformation of the Middle Ages into a "Renaissance" extending itself further and further back into history was the logical result of this historiographical dissolution of the dualism of the Enlightenment; it can be understood as the rationalization of the irrational element in rationality's self-understanding, and thus as a logical result of the Enlightenment itself. But the consequence was that the modern age seemed to lose its definition as an epoch, and thus also the legitimacy of its claim to have led man into a new and final phase of self-possession and self-realization.

An attempt to comprehend the structure of the change of epoch with rational categories was not made by the Enlightenment and up to the present has remained stuck fast in the dilemma of nominalism versus realism in interpreting the validity of the concept of an epoch. Wölfflin's resigned attitude in his *Kunstgeschichtliche Grundbegriffe* of 1915 is characteristic of the state of theory: "Transition is everything, and it is difficult to contradict one who regards history as an endless flow. For us, intellectual survival requires that we arrange the unlimitedness of events according to a few points of reference." A return to the self-interpretation of the Enlightenment, or merely to taking seriously the uniqueness of the character of the modern age as an epoch, as implied in that self-interpretation, is apparently impossible so long as even a vestige of this overwhelming "historicity" is operative in our attitude to the interpretation of history. But must the validity of the epochal category and the rationality of historiographical object definition be in conflict with one another? The answer is plainly yes, so long as the logic of continuity takes as its only alternative the constancy of what "was there all along" or preformation extending as far back as documentation is possible.

The insight that all logic, both historically and systematically, is based on structures of dialogue has not yet been brought to bear in the construction of historical categories. If the modern age was not the monologue, beginning at point zero, of the absolute subject—as it pictures itself—but rather the system of efforts to answer in a new context questions that were posed to man in the Middle Ages, then this would entail new standards for interpreting what does in fact function as an answer to a question but does not represent itself as such an answer and may even conceal the fact that that is what it is. Every occurrence [*Ereignis*], in the widest sense of the term, is characterized by "correspondence"; it responds to a question, a challenge, a discomfort; it bridges over an inconsistency, relaxes a tension, or occupies a vacant position. In a cartoon by Jean Effel in *L'Express*, De Gaulle was pictured opening a press conference with the remark, "Gentlemen! Now will you please give me the questions to my answers!" Something along those lines would serve to describe the procedure that would have to be employed in interpreting the logic of a historical epoch in relation to the one preceding it. Nietzsche understood the modern age as the result of intellectual pressure under which man had formerly lived, not at all, however, as the mere Cartesian "*jetter par terre*" [throwing down] of everything inherited, which was supposed to make possible the planned new construction, but rather as a specific correlate precisely focused on the prior demand and challenge. It is not difficult to eliminate the biologism of the idea of "training" from what Nietzsche says in *Beyond Good and Evil*:

> The long unfreedom of the spirit, the mistrustful constraints in the communicability of thoughts, the discipline thinkers imposed on themselves to think within the directions laid down by a church or court, or under Aristotelian presuppositions, the long spiritual will to interpret all events under a Christian schema and

to rediscover and justify the Christian God in every accident—all this, however
forced, capricious, hard, gruesome, and anti-rational, has shown itself to be the
means through which the European spirit has been trained to strength, ruthless
curiosity, and subtle mobility . . .[1]

If one translated character here into function and "strength" into argumentation, one arrives at a schema for a historiographical relation in which the
Middle Ages have lost the historiographical contingency vis-à-vis the modern age that they had in the Enlightenment's conception of history, their
arbitrariness as an annoying episode of confusion and obscurity in the text
of history, and in which they gain their historiographical relevance precisely
from their being able to provide the key to the sum of the requirements, as
the implicit satisfaction of which (whether the supposed or the actual satisfaction is of no concern) the modern age organized itself.

It is not an accident that Nietzsche names "ruthless curiosity" as one of
the epochal characteristics of the modern age that can only be understood
in its specificity and energy by reference to the passage through the Middle Ages. The "theoretical attitude" may be a constant of European history
since the awakening of the Ionians' interest in nature—a presupposition
that Edmund Husserl made basic to his conception of a teleology of this
history—but this attitude could only take on the explicitness of insistence
on the will and the right to theoretical curiosity after contradiction, restriction, competition, and the exclusiveness of other essential human interests
had been set up in opposition to it. After the Middle Ages, theory could no
longer be a simple continuation of the theoretical ideal of antiquity, as if a
mere disturbance of some centuries in length had intervened. Not only did
pent-up energy have to be let out through curiosity once it was rehabili-
tated, a kind of energy that deprived the ancient ideal's contemplative repose
of the qualities precisely of repose and calm; the medieval reservations had
also defined and given direction to a concentration of the will to knowledge
and of interest in specific realms of objects. To understand this concentration as a continuation of antiquity is to participate in a misunderstanding
suggested by the unavoidable employment of traditional terminology and
sanctioning formulas.

Among Goethe's aphorisms and fragments on the history of science,
there is a sketch that attempts to display the epochs of the sciences systematically according to the spiritual powers of man engaged in them: Sensual-

ity and imagination are the basis of a first, childish phase of the cognitive interest, which expresses itself in the form of *poetic* and *superstitious* views; in the second phase, sensuality and understanding are the basis of an empirical world interpretation, which is typified by *curious* and *inquiring* individuals; dogmatism and pedantry are the characteristics of a third epoch, in which understanding and reason combine for purposes that are predominantly *didactic*; in the fourth and last epoch, reason and imagination enter into a constellation of the *ideal*, whose opposing poles are designated as *methodical* and *mystical*.[2] In this schema, the attributes of the individual epochs, which are equipped with polarizing signs, arise entirely from the changing interplay of human faculties. The tensions lie in the modes of expression of the epochs themselves; but there is no visible logic leading from tension to intensification, that is, for example, from the childish polarity between poetry and superstition in viewing the world to the empirical opening up of access to the world in the field between the negative pole of the curious and the positive pole of the inquiring attitude to the world. (The assignment of plus and minus signs is Goethe's.) The peculiarly unhistorical character of the anthropologically compartmentalized schema conceals the historical logic in which attitudes of faith and superstition arrive at their own stages of dogmatic pedantry and, by means of the appearance of systematic completion and stability, block the view of what could endanger the system.[3] But curiosity, the instinct of inquiry, empirical openness are awakened precisely by the tabooing coercion of the dogmatic system, which not only must deprive its adherents of certain questions and pretensions but grounds this renunciation on a particular appropriateness and serviceability of the system. The extent to which the impossibility of the system's failing to answer questions loses self-evidence and to which the meritoriousness of renunciation or the viciousness of boundary infringement require argument is not only an index of the remaining unsatisfied curiosity or of the turning tide of awakening dissatisfaction but also reacts on this in a stimulating, accentuating, tendency-promoting fashion. The still "medieval" concessions and restrictions with which the seventeenth century wants to secure a path and legitimacy for the cognitive appetite always operate at the same time as reflections on what still remained to be set free.

At the turn of the seventeenth century, theoretical curiosity gains typification, definition as a figure, wealth of gesture. In the poetic figure of Doctor

Faustus there is created a bearer of its transformations and of the progress in its vindication. The original figure of Faust, in Johann Spies's *Historia* of 1587, still embodies terror at the sinful cognitive appetite, which "took for itself eagle's wings and wanted to investigate all the foundations of heaven and earth." The English translator already moderated the epithets of moral reprehensibility, and Christopher Marlowe transformed the baseness of the cognitive drive that is ready to risk anything into tragic greatness. Damnation still remains the ultimate consequence, but doubt has set in whether the spirit, when it surrenders itself entirely to its most characteristic motive, can be sinful. The chorus closes the tragedy with the moral that one should contemplate Faust's downfall so as to be warned by his fate not "to wonder at unlawful things./Whose deepness doth entice such forward wits/To practice more than heavenly power permits" [in the German translation: "*Zu tun, was hie und da der Seele wenig nütz*," to do what sometimes is not good for the soul]. Lessing's Faust was to be the first to find salvation; Goethe's Faust also found it—but did salvation resolve the question that gave the figure its epochal significance?

In his preface to Wilhelm Müller's translation of Marlowe's *Faust* (1817), Achim von Arnim "reclaimed the freedom of the world, as it matures, to rework this material" and grounded his assumption that "not enough Fausts have yet been written" on the "enormous arrogance of which the sciences of our time in their ingenious development are guilty," without meaning to cast doubt on the effective power of Goethe's version: "The further the lust for science spreads, the higher grows the arrogance of the individuals who think they have accomplished something and then deify themselves, the more renunciation science demands, the more the taste for science spreads—the deeper will the earnest truth of Goethe's *Faust* be felt . . ."

When in 1940 Paul Valéry availed himself, in what was perhaps a final, insurpassable instance, of the right to render the story incarnate again, to assign to Faust and Mephistopheles their metamorphoses as *instruments de l'esprit universel* [instruments of the universal spirit] in history, to recognize in them the humanity and inhumanity of the altered world—the result was a comedy with magic tricks. Why so? Because it was no longer Faust who needed to be tempted but his tempter: The process of knowledge itself had surpassed everything that could make magic enticing. The great gesture of curiosity has lost its scope when pointers indicate the instant at which to press a button.

As Christopher Marlowe wrote his *Faust*, in 1588, Giordano Bruno had already begun to pursue to the end his path of triumphant defiance on behalf of *curiositas*. At least in what he believes man capable of knowing about the world, he is the real Faust figure of the century, in distance from the Middle Ages far in advance of his poetic colleague. In the first dialogue of the *Ash Wednesday Supper*, he celebrates the ark of knowledge, his knowledge, as the penetration of heaven and overstepping the bounds of the world, as the opening of the prison where truth was held in custody and the exposure of concealed nature—in other words, as the *coup de main* of the cognitive drive against its medieval enclosure and limitation. Augustine's suspicion that behind astronomy stands the striving to elevate oneself to heaven by one's own power seems to be confirmed in Bruno: "*Non altrimente calcamo la stella, e siamo compresi noi dal cielo, che essi loro*" [Just as we tread our star and are contained in our heaven, so are they] declares the introduction to his dialogues on the infinite universe. Knowledge of nature and possession of happiness are identical; but the Aristotelian formula regarding man's cognitive drive, which Giordano Bruno places at the beginning of his interpretation of Aristotle's physics, is now directed precisely against the closed universe of Aristotle and the Middle Ages; it has become a formula of liberation rather than justification. This liberating cognitive striving is no longer the common nature of all men—otherwise Bruno would not feel the loneliness and forlornness of his fate—but rather the business of the few whom he can pull along with him and draw into the ecstatic flight of curiosity: "*O curiosi ingegni, / Peregrinate il mondo, / Cercate tutti i numerosi regni!*" [Oh curious spirits, / Travel around the world, / Investigate all the numerous kingdoms!].

Francis Bacon defends theoretical curiosity entirely differently, more indolently, with juristic tricks and shrewd twists of hallowed arguments. Just as he treats the process of nature in juristic categories "exactly like a civil or criminal matter" (as Liebig observed),[4] so also for him man's relation to nature, in the entire range—whose breadth he was the first to perceive—over which theoretical and technical mastery can be achieved, emerged as a question of legal rights. The distinction, which is usually understood as merely a result of shifting terminology, between eternal laws of nature (*leges aeternae*) and their usual course (*cursus communis*), is established in strict accordance with the distinction between codified law and common law and serves precisely to delimit the possibilities of human intervention in nature.

A parallel distinction appears in the roles assigned to metaphysics and physics. The former has as its object the unalterable law beyond man's influence; the latter comprises all knowledge of the operative and material causes that man can transpose in order to influence given states of affairs.[5] Thus Bacon recognizes no thoroughgoing lawful determination of nature, and I doubt whether today one still correctly understands his famous proposition that nature can only be mastered if one obeys her in the way one immediately assumes one can understand it, that is, by relating obedience and mastery to one and the same aspect of law.

The idea of an essential human right to knowledge, a right that has to be recovered, dominates Bacon's *Novum organum* (1620). The preface deals with the stagnant condition of the contemporary sciences and promises to open a new and hitherto unknown path for the human spirit and to provide it with resources so that it can "exercise its right with respect to nature."[6] Mankind's pretension to science is grounded in a divinely bestowed legal title, whose full power should now be exercised to the extent permitted by reason and "sound religion."[7] Bacon sees the satisfaction of the human appetite for knowledge with the achievements of the ancient world as a result not of conscious self-restriction in view of supposed limit settings but rather of an illusory underestimation of one's own powers and means, which at the same time can be described as an overestimation of what has been achieved. The pillars of Hercules, which are presented on the title page of the *Instauratio magna* as already being transcended by shipping traffic, are indeed a fateful boundary (*columnae fatales*)—but rather than representing a divine warning against hubris they represent the discouragement of desire and hope by myth.

Bacon avoids burdening the medieval sanction of the traditional world frame with the responsibility for man having allowed his possibilities to remain unexploited; as I will show, he must keep the theological premises unburdened so as to be able to derive from them a new legitimacy. Thus it is not religious humility and theological taboo that have brought about the great stagnation but rather man's error regarding the scope of his potential power; imagined riches are one of the main causes of poverty: "*Opinio copiae inter maximas causas inopiae.*" False trust in the world in the present makes one neglect the available sources of assistance for the future. False trust in the world—that is the dominant concern in Bacon's momentous exclusion of the teleological view of nature. But it is no longer the hidden and incomprehensibly sovereign God of nature Who denies man insight and intervention in nature but rather the historical indolence of man himself, who fails to recognize the goals of his interaction with nature and prematurely blocks the path of progress for himself by his faith in a special favor accorded to him by nature. The positions of hubris and an appropriate assessment of one's worth are interchanged in Bacon's system; men's carelessness and indolence (*socordia et inertia*) made them accept and treat as an agreeable authority the self-confidence and presumption of an arrogant spirit (of Aristotle), qualities that saved them the trouble of further investigations.[8]

This picture of the human spirit's historical indolence and unwillingness to progress presupposes of course an altered concept of theory itself, which no longer is the reposeful and bliss-conferring contemplation of things that present themselves—as the ancient world had regarded it—but rather is understood as work and a test of strength. It is no longer sufficient to draw the individual object into the focus of observation and, so to speak, expect it to give evidence about itself; only efforts to alter reality contribute to its explanation, and the patience of reposeful contemplation is useless.[9] Here also, without any notice being taken of the telescope and the microscope, the relevance of the invisible is brought into play, not, however, as the relevance of what is too small or too distant but rather as the way in which nature's constitutive elements and powers are hidden behind the self-presentation of its surface.

The novelty of the method of induction is the way in which it directs one's attention primarily away from the object of interest itself. Reason left to itself, the *intellectus sibi permissus*, is impotent because it is consumed in the momentariness of supposed evidence that Bacon calls *anticipatio*. This evidence penetrates the intellect in the moment and fills up the imagination, whereas reason that is equipped with and controlled by method is directed at a factual material that has not yet been organized, a material that does not constitute a pregiven coherent context but must first be brought into one and whose interpretation only becomes possible in such a context.[10] The critical side of this concept of knowledge is directed against the concept of reality as evidence in the present, and the new scheme rests entirely on the concept of reality as experimental consistency, in which the true nature of things—like that of the citizen in the state—shows itself only when they are withdrawn from their natural condition and exposed to, as it were, artificial disorder: *cum quis in perturbatione ponitur.*[11] Here the turning is described, from which Galileo extracted greater consequences than Bacon, when he turned his back on "natural" nature and out of curiosity had himself shown the arsenal of Venice, where the "artificial" nature embodied in the mechanisms of the instruments of war gave up its secrets. Bacon's ideal of knowledge still stands closer to the medieval undercurrent of magic than is the case with Galileo's physics. On the other hand, Galileo's confidence in the demonstrative power of optically strengthened observation, by means of the telescope, is more ancient in its concept of reality than our mythicized version of the scene makes it appear. Bacon's concept of *indirect* experience indicates the new path more explicitly here. The magic *habitus* is also in large part stylization, intoxication with metaphor. Bacon's rejection of mathematics as the medium in which to formulate the knowledge of nature is connected with his definition of the paradisaic condition as mastery by means of the word.[12]

Bacon wrote down the clearest outline of his justification of human curiosity by rearranging the biblical paradise as a utopian goal of human history as early as 1603 in the fragment entitled *Valerius terminus*, which, however, was only published posthumously in 1734. It is the earliest form of the projected systematics of his major works. The first chapter of the fragment carries the heading "Of the Limits and Ends of Knowledge." This problem is immediately traced back to two analogous events of theological prehistory: the fall of the angels and the fall of man. The relation between the two events lies in the interchange of the motivations appropriate to the behavior of each species, motivations that thereby become culpable: The angels, destined for the pure contemplation of divine truth, aspired to power; men, equipped in their paradisaic condition with power over nature, aspired to the pure and hidden knowledge.[13]

At the same time, the relation between knowledge and power is prefigured in the mythical reversal. The angel of light, endowed with pure knowledge but destined to serve the highest power, reaches for the highest power itself; man, equipped with power over everything created, but not partaking of the deepest secrets of the divinity, confined as spirit in the body, succumbs to the temptation of the light and liberty of knowledge.[14] Now he is endangered by both things, the lost sovereignty over reality and the longing for purity of knowledge, which for this being is as dangerous as stumbling upon a spring whose torrents of water gush forth without any prepared course. But the sacrilegiously arrogated knowledge is not that of things and nature; rather it is "moral knowledge," the ability to distinguish between good and evil. This alone is God's secret, which He wanted to reveal but not to expose to an autonomous grasp. By taking the biblical "tree of knowledge" literally, Bacon reserves the realm of morality for religion but gains nature as the harmless object of an inquiry that wants to and may regain for man his lost paradisaic dominion.

There is no hint of any doubt whether the regaining of the paradisaic power is compatible with the divine reservation in favor of revelation when the latter remains restricted to morality, since after all the lost paradise evidently had no need of morality and its discovery immediately entailed banishment. Bacon is not concerned with such inconsistency; it is enough for him that knowledge of nature was not the temptation that brought man to his fall.[15] If God appeared to have His secrets in nature as well, then that meant no prohibition but rather something like the divine majesty's delight in hiding His works, in the manner of innocent children's games, in order to let them ultimately be discovered nonetheless.[16] For Bacon, the great world hide-and-seek of the hidden God of late-medieval nominalism, which Descartes intensified into the suspicion of the universal deception of a *Dieu trompeur* [deceiving God] and sought to break through by grounding all certainty on absolute subjectivity, has exactly the innocence of a game laid out with the goal of eventual discovery and solution and free of any suggestion of jealousy of man's insight into the secret of the creation.

The idea is foreign to Bacon that human knowledge might be an auxiliary construction in place of what is unknown, might be hypothesis and mere probability—a consideration that the seventeenth century used over and over again to conceal its cognitive pretension. For Bacon, the human spirit is a mirror capable of containing the image of the universe.[17] The representation of this universe is determined by the idea of natural law ("law of nature"). The interpretation of this expression is still entirely bound to its metaphorical substratum, to the analogy of political law. This confines the idea within the horizon of the political thought of the time: The citizen does not enjoy insight into every aspect of the will of the ruler, but events and changes enable him to infer the reason that underlies the whole.[18]

Bacon loves still another set of metaphors, those of organic growth. Here as elsewhere they have the function of justifying the continuation of a formative process, once begun, toward a totality. Knowledge is "a plant of God's own planting," and its development to blossom and fruit, laid out in its inner principle, is "appointed to this autumn of the world," the harvesttime that Bacon believes has arrived in his century, as the time when his philosophy is "due," and which announces itself in the opening of the world by seafaring and trade and in the awakening of knowledge. Religion, so he argues, should support and promote the knowledge of nature, since its neglect is an offense to God's majesty, as though, perhaps, one were to judge the accomplishments of an outstanding jeweler only by what he lays out in his show window. But in everything the lost paradise remains the regulative idea of cognition; knowledge continues to be functionalized for power, for "the benefit and relief of the state and society of man"; otherwise it degenerates and "maketh the mind of man to swell." Here "the pleasure of curiosity" appears once more in its medieval signification; but the expression describes not a particular variety and objective reference of knowledge itself but rather a standstill in the pursuit of knowledge, a forgetfulness of its original purpose, a refusal of its potential for the recovery of paradise.[19] Knowledge that only satisfies itself fails to serve its organic purpose and takes on the characteristics of a sexual vice that produces no offspring.[20]

The voyage beyond the Pillars of Hercules has lost its adventurousness and no longer aims only into the enticing indefiniteness of the ocean; the certainty of finding *terra incognita* on the other side of the ocean justifies the departure; indeed it renders it criminal to linger in the land-locked sea of what is known. The extent to which we are certain that we do not yet know and rule nature, that we are not yet near the lost paradise again, guarantees the future of new discoveries. The increment to the known world that had already come about in Bacon's time contracts, in the great framework of this speculative picture of history, into a mere beginning, a symptom of possibilities.[21] Pure knowledge, whose idea coincides for Bacon with the ancient idea of theory, appears to him as an attitude of inescapable resignation because it has no motive for its progress but rather dwells persistently on each of its phenomena and loses itself in admiration of it.[22] *Curiositas* has become a worldly "sin," the indolence of theory in theory itself, a failure in the extensiveness of the cognitive pretension as a result of its intensiveness. The premature assertion, in the tradition, of boundaries to knowledge and capability was for Bacon the result of this ideal of pure theory.[23]

This position of Bacon's entails a leveling off and homogenization of the world of objects. There are no preferred and no unworthy objects of theory. The teleological interpretation of knowledge itself excludes any teleological interpretation of its objects. What an object "imports" first emerges only when it functions as a source of evidence to be assimilated by method, that is, not in what it is but in what it makes possible. Here a new concept of the "purity" of theory is forming itself, one that no longer has anything to do with the ancient ideal but rather points to what we nowadays call "basic research," where we exclude only *predefined* purposes, but we certainly do assume that theoretical results themselves give rise to possible goals, open up the path to applications. The general goal of the reattainment of paradise cannot be made specific in the form of fixed goals for the individual component processes. Here also the image of the contemporary voyages of discovery dominates Bacon's thought. No assumption of an unknown goal guides the ship's voyage; rather the compass enables one to hold to a path on which, in the field of the unknown, new land will eventually appear. Which goals are attainable is something that emerges from and along the paths that are found; method is the unspecific potentiality of knowledge, the security of holding to the chosen direction and the clarification of the possibilities inherent in it.

The systematic topography of the paths guarantees that the accidents of things coming to light ultimately lead to a universal acquaintance with the world. The discovery of the compass enables one to imagine a net of coordinates laid over reality and independent of its structures, in which the unsuspected can be sought and arranged. So much had remained concealed from the human spirit throughout many centuries and was discovered neither by philosophy nor by the faculty of reason but rather by accident and favorable opportunity, because it was all too different and distant from what was familiar, so that no preconception (*praenotio aliqua*) could lead one to it. Thus one may hope that nature conceals in its womb still more, of greater importance, which lies entirely outside the familiar paths of the power of imagination (*extra vias phantasiae*) and which one can only be sure of finding through the systematization of accident.[24] The tendency of Bacon's method is to set the human mind in controlled motion. Where nothing is undertaken, nothing can be achieved; and to expect representations of the goal where imagination and conception fail is to be misled, with standstill as a result.

The mythical construction of the fall of the angels and men and of the immanent restoration of paradise, which serves Bacon as a justification of human curiosity with respect to nature, presupposes a "politomorphic" conception of the world, that is, a conception in which natural laws are interpreted as decreed by divine volition and the role of created things is defined in the plan of creation in terms of service and power. What lies open and what remains hidden, what results in good or evil, is determined by this quasi-political state. Commandment and law, which were promulgated over nature verbally and are carried out according to the word, also have the word as their appropriate medium of knowledge. The determined antithesis of this position is constituted by the metaphysics of the mathematization of natural science. It proceeds from the impossibility in principle of secrecy in nature and things withheld from knowledge, to the extent that mathematical regularities are implemented in nature.

The mathematically comprehensible law of nature is not a decree of divine volition but rather the essence of necessity, in which divine and human spirit possess the same evidence, which as such excludes the withholding that makes things inaccessible. Kepler and Galileo gave the most compelling expression to this idea, more compelling than Descartes, because Descartes had to take the indirect path of securing certainty by means of the divine guarantee even for mathematics. The human spirit will correctly assess its powers, Kepler writes to Mästlin on April 19, 1597, only when it understands "that God, Who founded everything in the world according to the norm of quantity, also gave man a mind that can grasp these norms. For like the eye for color, the ear for sounds, so man's mind is not meant for the knowledge of whatever arbitrarily chosen things, but for that of magnitudes; it understands something the more correctly, the more it approaches pure quantities, as the origin of the thing."

The basic nominalist idea of mathematics as a constructive makeshift of knowledge over against the pure heterogeneousness of the world is thus given up and a form of justification for curiosity is found, which at bottom is the absence of any need for justification. Though Kepler may still give the idea the pious form of a derivation from man's creation in God's image, this is nevertheless inessential for the stringency of the argument; that we "gain a share in his own thoughts," as Kepler writes to Herwart on April 9/10, 1599, depends on the essential "openness" [*Öffentlichkeit*] of these thoughts themselves in their mathematical structure. In the knowledge of numbers and magnitudes, Kepler goes on, "our knowing, if piety permits one to say so, is of the same species as the divine knowing, at least insofar as we are able to grasp something of the latter in this mortal life. Only fools fear that we thereby make man into a god; for God's decrees are unfathomable, but not His corporeal works." Kepler has not yet loosed this idea from its medieval embedding in the idea of the preference given to man by providence; Leibniz was to be the first to accomplish this, by making the principle of sufficient reason into the criterion by which to verify the divine plan of creation objectively and basing on this the impossibility that the nature of the *mathesis divina* [divine mathematizing] should be arcane.[25]

This is no "secularization" of man's having been created in God's image. The function of the thought emerges naked and undisguised and makes its historical derivation a matter of indifference: Knowledge has no need of justification; it justifies itself; it does not owe thanks for itself to God; it no longer has any tinge of illumination or graciously permitted participation but rests in its own evidence, from which God and man cannot escape. The Middle Ages of High Scholasticism had seen man's relation to reality as a triangular relation mediated by the divinity. Cognitive certainty was possible because God guaranteed man's participation in His creative rationality when He brought him into the fellowship of His world idea and wanted to furnish him, according to the measure of His grace, with insight into the conception of nature. Any autonomous step beyond this conception strained the relation of dependence and the debt of thanks. This triangular relation is now dissolved; human knowledge is commensurable with divine knowledge, on the basis, in fact, of the object itself and its necessity. Reality has its authentic, obligatory rationality and no longer has need of a guarantee of its adequate accessibility. The problematic of theoretical curiosity, which had depended on the idea of the world as a demonstration of divine power, and of human stupor as the corresponding effect, is paralyzed by the idea that knowledge is not a pretension to what is unfathomable but rather the laying open of necessity.

From the point of view of the stage that the argument was to reach with Leibniz, the position that Galileo occupies in developing it, and the burden that he had to bear in opposing the theological position, becomes clearer. At the end of the "First Day" of the *Dialogue on the World Systems*, the question of the existence of living beings on the moon is discussed. The human power of imagination, Salviati decides, is incapable of imagining such beings, since the riches of nature and the omnipotence of the Creator lead one to expect that they would be utterly different from the ones known to us. Sagredo agrees with him that it would be the highest audacity (*estrema temerità*) to make the human power of imagination the criterion of what can exist in nature. Only a complete lack of all knowledge would lead to the vain presumption (*vana presunzione*) of wanting to understand the whole. *Docta ignorantia* [learned ignorance], the Socratic knowledge of ignorance, is praised as the effect of true cognition; no state of knowledge attainable in fact could diminish the difference between human knowledge and divine knowledge in its infinity: "*Il saper divino esser infinite volte infinito*" [Divine knowledge is infinitely times infinite]. So far this could have been taken from a medieval treatise.

Galileo's artful dialectic, however, consists in the fact that he puts in the mouth of the conservative figure of the Scholastic in his dialogue, Simplicio, the remark that gives the argumentation its turning. If this were how things lay with regard to human knowledge, Simplicio says, one would have to admit that even nature had not understood how to produce an intellect that understands.[26] Salviati praises the acuteness of the otherwise dull mind, and this always means that he can catch "the Middle Ages" in an inconsistency. As a matter of fact everything that one had heretofore admitted to be a weakness of the human spirit had been related to the extent of knowledge. But if one considered its intensity, that is to say, the degree of certainty of an individual proposition, then one would have to admit that the human intellect can grasp some truths just as perfectly, and possess certainty regarding them just as unconditional, as any that nature itself could possess.[27] It is true that the divine spirit can grasp infinitely many more mathematical truths than the human spirit, namely, all of them, but knowledge of the few is equivalent in objective certainty (*certezza obiettiva*) to divine knowledge, even if the species of the cognitive act may be different.

And now follows the crucial sentence, which brings the whole train of thought into relief against the background of what was still medievally thinkable: The human knowledge of mathematical truths conceives them in their necessity, beyond which there can be no higher level of certainty.[28] These statements are common property and free from the suspicion of presumptuousness or daring (*lontane da ogni ombra di temeritá o d'ardire*). The dialogue has traversed the obligatory stretch of humility and earned its license to enhance the dignity and accomplishment of the human spirit. That nullity is not so null as to prevent the "First Day" of the dialogue from ending with a *laudatio* [eulogy] of the human spirit's acuity and discoveries.[29]

When Galileo had completed the *Dialogue on the World Systems*, he was once again, and forcibly, confronted with the question of the legitimacy of the cognitive will. The Roman censor imposed on him the stipulation that the theses of the work—especially the explanation of the tides by the movement of the earth, in the "Fourth Day," a movement that is the physical and realistic correlate of the Copernicanism that was permitted only as an astronomical hypothesis—should be put under the proviso of the infinite possibilities open to omnipotence. One could say that this proviso of sovereignty forced upon Galileo did not affect the core of his epistemology because to all intents and purposes the whole dialogue made no use of the technique of mathematical astronomy; that is, it did not even touch that highest intensive dignity of the intellect, for which Galileo had postulated an evidence common to God and man. The irony of the situation lies in the fact that the Pope had informed Galileo, through the inquisitor of Florence, that the mathematical presentation of the Copernican doctrine remained entirely open to him. But Galileo gathers together all the materials that he himself, according to his classification, would have to ascribe to the *extensive* accomplishment of knowledge and thus render suspect of the weakness that is proper to knowledge wherever it does not become mathematical.

Admittedly, where the Copernican matter is concerned, Galileo's own criteria leave him in the lurch. The formula that Galileo had to work into the text does not contain, as has been asserted, something like Urban VIII's personal countertheory to Galileo's explanation of the tides; rather it is an "antitheory" altogether, opposed to physical theory of any kind. In it the difference between the thought of the spent Middle Ages and the new pretension of the scientific explanation of nature comes undisguisedly to light. Offense was to be taken at the fact that Galileo placed this vexatious formula in the mouth of Simplicio, that is, of the figure in the dialogue who is the loser in the end. There is, he says, always present in his mind a teaching that was once given him by a highly learned and highly placed person and that he regards as unshakeably well established, namely, that God can also produce a phenomenon in nature in a different way than is made to appear plausible by a particular explanation. Consequently it would be inadmissible daring (*soverchia arditezza*) to want to narrow down and commit divine power and wisdom to a particular idea by asserting that single explanation to be true.

This general proviso throws man back, as far as he is concerned with theoretical truth, to a hopeless position. For him the world would continue to have an unperspicuous structure, whose laws had to remain unknown to him and for which any theoretical exertion would stand under the threat of the revocation of the condition of its possibility. Of course for the theologian Maffeo Barberini, who sat upon the papal chair, God *could* be an uncertainty factor in knowledge, since at the same time He offered man His revelation as the sole salvation-bringing certainty and had the right not to allow its uniqueness to be leveled off or supplanted by other supposed certainties.

It is evident that what Galileo had to deal with here was the charge—which had now reached the stage of "taking measures"—embodied in the verdict against *curiositas*. The discipline with which Galileo makes Salviati answer that this is an admirable and truly angelic doctrine allows him a new bit of dialectic: It is perfectly in harmony with this to award us permission to *investigate* the construction of the world, even if we may not pretend *fully to understand* the work, as it comes forth from the divine hands, since otherwise the activity of the human spirit might perhaps grow weak and be exhausted.[30] Theoretical curiosity, it seems to be argued here, has its economy between the futility to which the omnipotence proviso means to relegate it and the definitiveness in which belief in the completed possession of knowledge would fix it. Galileo's resistance to the omnipotence proviso suggests that it is only and especially the movement of the intellect, as progress in its understanding and its formulation of problems, that guarantees consciousness of the finitude of knowledge over against the infinitude of what, though it is never irreducible, is nevertheless kept in reserve at any given time, whereas the mere appeal to the infinitude of the intrinsically noncontradictory possibilities open to omnipotence destroys any consciousness of a relation between what is already known and what is still to be known and thrusts reason back into the indifference of resignation. The project of infinite progress does indeed direct itself against reason's theological resignation but against reflection on its finitude, which can only be experienced precisely in the accomplishment of its possibilities.

The role that the progress of knowledge begins to play in the justification of curiosity stands out in the negative at the close of the *Dialogo*: It is not what the progress of knowledge has already yielded and is yielding now that legitimates the cognitive drive that keeps it in motion but rather the function for consciousness of what lies before it at any given time, which gives everything that has been achieved the mark of finitude and provisionality. "Sir, my science is greedy for knowledge!" says Bertolt Brecht's Galileo in bargaining with the curator of the University of Padua, who would like to divert his wish for income to the more lucrative provision of private tuition. The declaration that Brecht puts in Galileo's mouth here objectivizes curiosity and makes it into a mark of the imperfect status of science; this corresponds to the objectivity of the conflict of systems in which Galileo and Urban VIII are the exponents. But Brecht does not stick consistently to this objectivization of curiosity. He writes in his *Notes on the "Life of Galileo"*:

> The inquisitive drive, a social phenomenon scarcely less lustful or dictatorial than the procreative drive, directs Galileo into this dangerous territory, drives him into painful conflict with his intense wishes for other satisfactions. He lifts the telescope to the stars and delivers himself over to torture. In the end, he practices his science like a vice, secretly, probably with pangs of conscience. In view of such a situation, one can scarcely be bent on either only praising Galileo or only damning him.[31]

Such a late reintroduction of theoretical curiosity into the catalog of vices has, of course, its new premises: Where it converts itself, as motive power, into science, it gives this science an imprint of "purity" and apragmatic disregard for consequences that makes it appear just as questionable from the humane or social point of view as it was under the primacy of the exclusiveness assigned to the question of salvation by theology. Galileo, writes Brecht,

> enriched astronomy and physics by simultaneously robbing these sciences of the greater part of their social meaning . . . Galileo's crime can be regarded as the "original sin" of the modern natural sciences. From the new astronomy, which deeply interested a new class, the bourgeoisie, since it afforded assistance to the revolutionary social currents of the time, he made a sharply circumscribed specialized science, which of course precisely on account of its "purity," that is, its indifference to the mode of production, could develop with relative freedom

from disturbance. The atom bomb, as both a technical and a social phenomenon, is the classical end product of his scientific accomplishment and his social failure.

The difference between the two statements about Galileo—the one in the play and the one in the notes on the play—gives us a key to the new shape of the problem of theoretical curiosity in the seventeenth century. In my opinion, the most precise comprehension of the actual state of affairs is extracted in the short statement to the Paduan curator: Curiosity is not only no longer able to be one of the vices of the individual in need of redemption; it has already separated itself from the structure of personality, from the psychic motive forces, and has become the mark of the hectic unrest of the scientific process itself.

Admittedly this corresponds neither to the picture that Galileo made for himself of his own type of mental eagerness nor to what his contemporaries saw in him. The digressiveness that dominates the style of both of his great dialogues and is so distant from the linear methodology that Descartes was to project for the new idea of science indulges, with emphatic unconcern for system, in the psychological attraction of each new *curiositá* [curiosity] as almost an isolated quality of the objects themselves. New truths are found off the side of the direct path of what method anticipates, by seizing accidental opportunities, by being ready to drop the thread of principles already established.[32] The opposite type is represented by the people who could not be curious enough to look through the telescope because they believed themselves to know too accurately already what one could not see with it. When in the "Third Day" of the *Dialogue on the World Systems* the discussion comes to William Gilbert's theory of magnetism, Galileo makes Salviati say that against the authority of the inherited conceptions, only a curiosity comparable to his own (*una curiositá simile alla mia*) and the suspicion of an infinite reserve of unknown things in nature[33] could maintain inner freedom and openness toward what is new. Martin Horky, the pamphleteer against the moons of Jupiter who was put forward by Magini, already placed Galileo's eye trouble, which was ultimately to lead to blindness, in an infamous relation with his curiosity: ". . . *optici nervi, quia nimis curiose et pompose scrupula circa Jovem observavit, rupti* . . ." [. . . optical nerves, which have observed Jupiter with too much curious and ostentatious scrutiny, broken . . .]. And

Galileo's first biographer, Vincenzio Viviani, made *filosofica curiositá* [philosophical curiosity] the key term in his characterization of his teacher.

But the crucial objection—precisely because it was no longer medieval—against curiosity as the power that was definitive for Galileo's style of inquiry was first raised and could only have been raised by Descartes, who for his part gave the new science not indeed its aims and contents but certainly its form and agenda. He writes regarding Galileo to Mersenne, "His error is that he continually disgresses and never stops to expound his material thoroughly, which shows that he never examined it in an orderly fashion and that without considering the primary causes of nature, he merely sought the causes of some particular effects, so that he built without a foundation."[34] The motor quality of theoretical curiosity appears in Descartes to be threatened by objective irritations, to which it all too easily succumbs, by being made to forget the basic and presuppositional questions, the foundations and critical reinforcements, to which a thinker of the Cartesian type devotes himself entirely. Here there arises a sort of intrascientific morality, a rigorism of systematic logic, to which the unbridled appetite for knowledge is bound to be suspect.

With decisiveness equal to that with which he excluded teleology from the canon of possible questions directed at nature, Descartes gave human knowledge the teleological character of a strenuous exertion, united by the Method, toward the attainment of the definitive morality, which as the epitome of materially appropriate behavior in the world presupposes the perfection of factual knowledge. The employment of the words *curiosité* and *curieux* [curiosity and curious person] in Descartes has neither pathos nor specificity; the rational goal of knowledge excludes any other justification of the energies that must be expended for its achievement. The *curieux* takes on the professional quality of the scholar, who is characterized more by the methodically secured or attainable possession of knowledge than by the elemental need for knowledge, even if something of the infamy of the *sciences curieuses* [curious sciences] may still adhere to the disciplines, such as anatomy or chemistry, that are now entering the orderly ranks of scholarship.[35] The predicate formerly attached to the solitary daring of individuals is made harmless as the designation of an interested public and as the characterization of the equally harmless activities of the collector and the amateur, which are carried along with the great train of scientific progress and protected by it.³⁶ Part of the process of the legitimation of theoretical curiosity is its effort to rediscover itself, or a preformation of itself, in the region where something new would in any case look for evidence to demonstrate that it is nothing new, before a consciousness that novelty is permissible had stabilized itself in history—namely, in antiquity.

The early historiography of philosophy—represented in its most lasting form by the widely read Jakob Brucker, who defined the picture of the philosophical tradition for, for example, Kant—already poses the question of what was special about the conditions under which something like "philosophy" could arise among the Greeks as the epitome of theoretical procedure—a question that was to be approached in many ways, for instance, that of expounding the disposition of the Greek language for the development of philosophical modes of thought. Here one must recognize to what an extent the beginnings of the historiography of philosophy depend directly on the testimony of the sources themselves, since the critical thinning out of the tradition, which begins with Bayle, affects the anecdotal material more than the study of doctrine; but it affects the latter even more than it does the statements, withdrawn from verification, about the origin and the motives of the theoretical attitude.

In the nominally short, actually seven-volume long *Kurtze Fragen aus der Philosophischen Historie* [Short Questions from the History of Philosophy], whose first volume was published in Ulm in 1731, Brucker too asks the question (formulated in the manner of a Scholastic *quaestio*): "What was the beginning of philosophy like, with the Greeks?" Brucker regularly answers such questions initially with a single, and therefore usually not very informative, sentence, as in this case: "Rather scanty and very obscure as well."³⁷ "Wide-awake" Greeks had gone to the Orient, and there had seen the "establishments" of people "who were considered learned"; and in the reverse direction, foreigners had settled in Greece and given instruction. The philosophy of the barbarians is characterized as *philosophia traditiva* [traditional philosophy], as a canon of learned answers, handed down unthinkingly, to established questions. The Greeks, however, were moved to reflection by this teaching. This version of the relation is important because it does not adopt the dogma that philosophy was an inheritance from the East but rather defines a new beginning in relation to this inheritance and not even as a result of unmediated wonder at nature and pressure the questions that it poses. It is remarkable how little use is made here of Greek philosophy's interpretation of itself, whether it be the designation of wonder as its original affect or the discovery of the supposed allegorical hidden meaning in mythology.

For Brucker, neither immediacy nor tradition can be the reason for this beginning. What is important for him is "the curiosity of the Greek nation." This is in harmony with the manner of speaking that prevailed in his century. Fontenelle will still insert this characterization of the Greeks into his Cartesian schema of the difference between *raison* [reason] and *esprit* [wit, intelligence], thus attributing to the origin of philosophy the quality that was supposed to have prevented it, until Descartes, from performing its plain function: "*Les Grecs en general avoient extremement de l'esprit, mais ils étoient fort legers, curieux, inquiets, incapable de se moderer sur rien; et pour dire tout ce que j'en pense, ils avoient tant d'esprit, que leur raison en soufroit un peu.*"[38] [The Greeks had a great deal of wit, but they were very fickle, curious, agitated, incapable of restraining themselves in any area; and, to be frank, they had so much wit that their reason suffered somewhat as a result.] Brucker does not see the quality of curiosity in terms of the antithesis between intellectual excitement and controlled rationality. Philosophy seems to him to be the result of a setting free of energies over against traditional doctrine. Consequently it is important that the motive of curiosity appears here in what was an unusual, if not unique association for the time: in association with favorable political circumstances. Under a "form of government in which anyone could think, say, and teach whatever he wanted to," the impulse of curiosity develops into science. Allegorical interpretation and esoteric doctrine become mere residues of the earlier traditional form of philosophy, "vestiges of the secret type of teaching."

So as to make more evident the contrast between Brucker's analysis or the origin of curiosity and the traditional harnessing together of the immediate impact of nature with theoretical excitation, I want to quote the very conventional formulations from J. G. Sulzer's *Gedanken über den Ursprung und die verschiedenen Bestimmungen der Wissenschaften und schönen Künste* [Thoughts on the Origin and the Different Purposes of the Sciences and the Fine Arts] (1757) on the beginning of the sciences:

> . . . an innocent curiosity and the desire to gain a complete acquaintance with the phenomena of nature provided the occasion for their genesis; and no doubt

this occurred later than the birth of the arts. Nature is a vast theater that presents amazing objects and events on all sides. Could men, once they were free of their initial struggle for sustenance and had gained some free time, for long regard this magnificent edifice of the world without thinking of the unseen power that brought forth such a thing and the dexterous hand that organized the parts into a whole? Could the old inhabitants of these fortunate regions, where a pure and peaceful air always leaves the heavens unobscured, for long view this marvellous vault in which so many stars sparkle, of which any one would be sufficient to fix our gaze, without asking themselves what all these lights may be?[39]

Nevertheless, for Sulzer too the disposition to such immediate admiration of nature is not present in everyone: "There are entire peoples in whom one finds not a trace of this curiosity, which is a mother of science. The stupid Hottentot and the miserable native of Greenland regard the wonders of nature with an amazing indifference." When Brucker comes back to this topic in the second volume of his work, in the "Additions and Improvements,"[40] the connection between curiosity and democracy becomes clearer: It destroys the form of the *philosophia traditiva* [traditional philosophy] as a wisdom guarded by priests, which conveys only dicta but not "the cause, the connection of the reasons with the conclusions . . . And just as soon as the ancient Greeks had laid aside bestiality and the savage state, they themselves began to reflect. . . ." In this process, the politician took the place of the priest: "The investigation of truth got away from the priests, and soon *politiki* [statesmen] applied themselves to it. . . ." This is very remarkable because it reverses, historically, the direction of the relation of foundation between philosophy and politics in the metaphysical tradition. This tradition had always seen the theoretical attitude as depending on leisure. Theoretical curiosity did not require any other public conditions than the negative one of freedom from the pressure of need; it did not require the possibility of satisfaction of the cognitive interest by virtue of the competence of everyone to inquire into everything.

The Question of Normality in the History of Biological Thought

Georges Canguilhem

Georges Canguilhem (1904–1995) was a preeminent philosopher and historian of science and medicine. Trained as a doctor as well as a philosopher, he developed an original style of inquiring into problems of biology, physiology, and medicine (among other topics) through the comparative historical study of theories, concepts, and objects. In this essay, Canguilhem explains the need for what he calls, drawing on the historian and biologist Emil Radl, an "idea of the biological": a philosophical warrant for distinguishing the scientific character of biology from scientific modes grounded in the laws of physics or chemistry. The need for such a concept derives from what he calls the normality of the organism: the fact that living beings assert their own life and health in relation to changing environments, under the constant threat of disease and death. Unlike biology, Canguilhem reminds us, the physical sciences have no need for a science of pathology. He offers a distinctly philosophical concept of the biological to equip scholars for inquiry into the history of the biological sciences.

The historian Emil Radl was surprised to discover that biologists and historians counted Galileo and Descartes among those to whom modern life science was indebted for its methods, when "neither of them was associated with any biological idea worth mentioning."[1] The Prague historian was clearly already critical of what later came to be known as reductionism, the first axioms of which were stated by the Vienna School. Radl was explicitly dissociating himself from a certain post-Darwinian philosophy of biology, in which positivist taboos and materialist injunctions were amalgamated into a laboratory rationalism identified with the political radicalism of the period. If life is nothing but matter, so much for the soul, immortality, and the power of the priests!

But what is a "biological idea?" Or, to put it another way, what is an "idea of the biological?" Can history provide an answer to such a question? And must one have an answer before one can consider the history of biology to be part of the history of science?

The history of science would surely fail of its goal if it did not succeed in representing the succession of attempts, impasses, and repetitions that resulted in the constitution of what the science today takes to be its object of interest. Unlike geometry and astronomy, terms that are more than two thousand years old, the term biology is not yet two hundred years old. When it was first proposed, geometry had long since ceased to [be] the science of figures that can be drawn with straightedge and compass, while astronomy had only recently expanded its scope of interest beyond the solar system. In both cases, the signifier of the scientific discipline remained the same, but the discipline in question had broken with its past. By contrast, the concept of biology was invented to characterize, in retrospect, a discipline that had not yet broken with its past.

The word *biology* occurs for the first time in Lamarck's *Hydrogeology* (1802). When he mentioned the word again, in the preface to his *Zoological Philosophy* (1809), it was in allusion to a treatise to be entitled *Biology* that he never actually wrote. Strikingly, this preface is concerned with general problems of animal organization "as one traverses their entire series from the most perfect to the most imperfect." The idea of a hierarchical series of animals, a "chain of being," indicates that the object of the new biology was the same as that of Aristotle's *Historia animalium* and *De partibus animalium*. Hence Lamarck's own invention—modification in the organs through force of habit and under the influence of changing environmental conditions— was explicitly intended to reestablish "the very order of nature" beyond the lacunae and discontinuities in the system of classification proposed by the naturalists, that is, to establish a clear progression and gradation in organization that could not be overlooked despite any "anomalies."

As for the other inventor of the term and concept of biology, G. R. Treviranus, the very title of the book he published in 1802, *Biologie oder Philosophie der Lebenden Natur für Naturforscher und Arzte* (volume 2 in a six-volume series, the last of which was published in 1822), indicates that he had no wish to separate or distinguish the naturalist from the physician as to their philosophical or general conception of the phenomena of life. Thus, at the turn of the nineteenth century, a new way of looking at the study of living things, which entailed a new logic, was in fact limited by the traditional association of the standpoint of the naturalist with that of the physician, that of the investigator with that of the healer. Cuvier in his *History of the Natural Sciences* (sixth lecture) emphasized Aristotle's debt as naturalist to Asclepiades. In the same spirit Charles Singer has written that since Hippocrates was anointed the "father of medicine," he might also be called the "father of biology."

Since the turn of the nineteenth century, however, definitions of biology's specific object have been purged of value-laden concepts such as perfection or imperfection, normality or abnormality. Therapeutic intentions, which once informed or, more accurately, deformed, the biologist's view of laboratory work, have since been limited to the applications of biological knowledge. Hence it would seem that the question of "normality" in the history of biology ought to be classed as a matter of historical rather than current interest. I shall attempt to prove the contrary. To that end, I direct the reader's attention to the end of the historical process. For contemporary biochemists, the functions of self-preservation, self-reproduction, and self-regulation are characteristic properties of microorganisms such as bacteria. The model often proposed by scientists themselves and not just by popularizers of their work is that of the "fully automated chemical factory."[2] The organic functions are acknowledged to be superior to their technological counterparts in reliability if not infallibility and in the existence of mechanisms for detecting and correcting reproductive errors or flaws. These facts make it reasonable to ask whether there is not some principle of thematic conservation at work in the historical constitution of biology. On this view, which contrasts with an idea of science elaborated by historians and philosophers in the era when physics dealt with macroscopic objects, biology is different from the other sciences, and the history of biology ought to reflect that fact in the questions it asks and the way in which it answers them. For the alleged principle of thematic conservation in the history of biology is perhaps only a reflection of the biologist's acceptance in one way or another of the indisputable fact that life, whatever form it may take, involves self-preservation by means of self-regulation. Might this be what Emil Radl meant by a "biological idea"? Without a doubt the road from Aristotle's entelechy to the biochemist's enzyme is long and winding. But is it really a road?

I

The fundamental concepts in Aristotle's definition of life are those of soul and organ. A living body is an animate and organized body. It is animate because it is organized. Its soul is in fact act, form, and end. "Suppose that the eye were an animal—sight would have been its soul . . . We must now extend our consideration from the 'parts' to the whole living body; for what the departmental sense is to the bodily part which is its organ, that the whole faculty of sense is to the whole sensitive body as such" (*De anima* II.1 412b18, J. A. Smith, translator). The organs are the instruments of the soul's ends. "The body too must somehow or other be made for the soul, and each part of it for some subordinate function, to which it is adapted" (*De partibus animalium* I.5 645b18, William Ogle, translator). It is impossible to overstate the influence of Aristotle's use of the term *organon* to designate a functional part (*morion*) of an animal or vegetal body such as a hand, beak, wing, root, or what have you. Until at least the end of the eighteenth century anatomy and physiology preserved, with all its ambiguities, a term that Aristotle borrowed from the lexicon of artisans and musicians, whose use indicated implicit or explicit acceptance of some sort of analogy between nature and art, life and technics.

As is well known, Aristotle conceived of nature and life as the art of arts, by which he meant a process teleological by its very nature, immanent, unpremeditated, and undeliberated—a process that every technique tends to imitate and that the art of medicine approaches most closely when it heals by applying to itself rules inspired by the idea of health, the telos and form of the living organism. Aristotle, a physician's son, thus subscribed to a biological naturalism that had affinities with the naturalism of Hippocrates.

Life's teleological process is not perfectly efficient and infallible, however. The existence of monsters (*De generatione animalium* IV.10) shows that nature does make mistakes, which can be explained in terms of matter's resistance to form. Forms or ends are not necessarily and universally exemplary; a certain deviation is tolerated. The form of an organism is expressed through a rough constancy; it is what the organism appears to be most of the time. Hence we can consider a form to be a norm, compared to which the exceptional can be characterized as abnormal.

Descartes contradicted Aristotle's propositions point by point. For him, nature was identical with the laws of motion and conservation. Every art, including medicine, was a kind of machine-building. Descartes preserved the anatomical and physiological concept of an organ but eliminated any distinction between organization and fabrication. A living body could serve as the model for an automaton or vice versa. Yet there was an ambiguity in this reversibility. The intention behind the construction of an automaton was to *copy* nature, but in the Cartesian theory of life the automaton served as an intelligible *equivalent* of nature. In Cartesian physics there is no room for an ontological difference between nature and art. "When a watch indicates the time by means of the wheels of which it is made, that is no less natural than when a particular tree or seed produces a particular fruit" (*Principes de la philosophie* IV.203).

It is not surprising that some historians of biology and medicine place Descartes in the same group as the Italian mathematician-physicians inspired by Galileo's mechanics and Santorius's medical statics. Yet other historians find this classification paradoxical, since it makes the reductionist enterprise a part of the history of biology even though its effect was to obliterate that science's distinctive subject matter—what I have been calling its specific object. To my mind, this rather scholastic distinction is unworkable, for it is based on an incomplete reading of the sources and inadequate attention to certain concepts. Descartes, I shall argue, did not succeed in winning adherents to his project or program because he was obliged to incorporate into his definition of life as an aspect of mechanics certain positive attributes that resisted assimilation to that view.

To begin with, the Cartesian watch is no less subject to the laws of mechanics if it tells the time incorrectly that if it tells the time correctly (*Méditations métaphysiques* VI). Similarly, it is no less natural for a man to be sick than to be healthy, and sickness is not a corruption of nature (ibid.). Yet the thirst that drives the victim of dropsy to drink is a "veritable error of nature," even though it is an effect of the substantial union of soul and body, whose sensations, such as thirst or pain, are statistically valid indicators of things or situations favorable or harmful "to the conservation of the human body when it is fully healthy" (ibid.). This idea is confirmed at the end of the *Conversations with Burman* (1648), in which the medicine of the physicians, not based on sound Cartesian mechanics, is denigrated and ridiculed in favor of a course of conduct amenable, as animals are, to the silent lessons of nature concerning "self-restitution." "Every man is capable of being his own physician."[3] Even for Descartes, self-preservation remains the primary distinctive characteristic of the living body. After Descartes's death (1650), the watch model was further elaborated by the addition of regulator devices, long after their invention by Huygens: the isochronous pendulum (1657) and the spiral spring (1675). When Lavoisier introduced the concept of "regulators of the animal machine" into the physiology of respiration and animal heat (in his *First Memoir on the Respiration of Animals*, 1789), Cartesian concepts were brought into line with Hippocratic intuition.

If the metaphorical concept of the "animal machine" could not totally conceal the characteristic feature of life for which the name "regulation" was suggested in the eighteenth century, the no less metaphorical concept of "animal economy," first put forward in 1640 in imitation of the notion of political economy (proposed in 1615), was explicitly intended to evoke the well-tempered structures and functions of the organized body. Like the domestic economy, the animal economy required wise government of a complex entity in order to promote the general welfare. In the history of physiology the idea of "animal economy" was responsible for a gradual shift from the notion of animal machine to the notion of organism over the course of the eighteenth century. Now, the concept of economy, like that of organ, can be traced back to Aristotle. The overlap of the two concepts is governed by some logic in the history of scientific ideas.

Evidence for widespread efforts to work out a new concept of life after 1650 can be seen in the proliferation of derivatives of the word *organ* in Latin, French, and English: organization, organized, organic, and organism, to name a few. These were used by both philosophers such as Gassendi, Locke, Leibniz, and Bossuet and physicians such as Duncan and Stahl. Undoubtedly it was Stahl who most stubbornly defended the irreducibility of the organism, that is, the idea that a certain *order* obtains in the relations of the parts of a mechanism to the whole (*De diversitate organismi et mecanismi*, 1706). A living body is both instrumented and instrumental. Its efficient

structure (*structura, constructio, ordinatio, distributio,* par. 19) reveals cooperation on the part of mediate or immediate agents. The material constitution
of the body is subject to rapid corruption. Stahl observes, however, that disease is an exceptional condition. Hence there must be some power of conservation, some immaterial power offering active resistance to decomposi-
tion, permanently at work in the bodies of living things. Self-preservation
of the organism is achieved as a result not of some mechanical but of natural
"autocracy" (*De autocratia naturae,* 1696).

The importance of Stahlian animism is not to be gauged by the refutation
of most of Stahl's ideas as physiology progressed. Assuming that the identification of enduring characteristics of organisms is a less fragile element
of Stahl's system than is the attribution of supposed causes of those characteristics, then Stahl left his mark on more than one nineteenth-century
biologist. He had his followers in Scotland and England (such as Robert
Whytt) and in Germany (such as Felix Platner) but most of all in France,
where the Stahlian school in Montpellier, led by T. de Bordeu and P.-J.
Barthez, inspired the work of Xavier Bichat. Claude Bernard's criticism of
Bichat's vitalism did not prevent him from acknowledging that his approach
to physiology owed as much to the reading of Bichat as to the example of
Magendie. Death, disease, and the capacity for recovery are the characteristics that distinguish life from mere existence. "Philosophical physicians
and naturalists have been deeply struck by the tendency of organized beings to reestablish their form . . . and thus to demonstrate their unity, their
morphological individuality" (*Leçons sur les phénomènes de la vie communs aux
animaux et aux végétaux,* 1st lecture, 1878). The causal theories proposed to
explain the regularity and consistency of organized phenomena mattered
less to Bernard than recognition of the fact of organization itself: "Hence in
the animate body there is an arrangement, a sort of ordering, that should not
be overlooked, for it is truly the most salient trait of living beings. I admit
that the idea of this arrangement is ill expressed by the term *force.* But *here
the word is of little importance; suffice it to say that the reality of the fact is beyond
dispute.*"[4]

It is not only the history of anatomy and physiology that begins with Aristotle but also the history of what was long called "natural history," including the classification of living things, their orderly arrangement in a table of
similarities and differences, study of their kinship through morphological

comparison, and, finally, study of the compatibility of different modes of existence. Natural history sought to explain the diversity of life forms able to coexist in a given environment. Linnaeus (1749) referred to this coexistence as the *oeconomia naturae*.

The dominant question in the history of natural history was that of *species*. What is the status of the set of determinants that distinguishes the wolf from the jackal, the buttercup from the rose? Is it nominal or real? The distinguishing features are enduring traits, yet they do admit variations and differences. Hence natural historians were obliged to investigate the conditions under which unity can subsist within diversity, and thus they were led to explain morphology in terms of genealogy, forms in terms of their mode of reproduction. Accordingly, they showed keen interest in questions of fertility and interfertility, hybridization and intersterility.

In the eighteenth century the status of species was the foremost problem of the naturalists, as can be seen most clearly of all in the work of Buffon and Linnaeus. The latter did not experience as much difficulty as the former in holding that the species were fixed at creation and perpetuated from generation to generation. Buffon attempted to resolve the problem with his theory of "internal molds" and "organic molecules." Organic molecules, he maintained, were indestructible; they survived the process of reproduction from generation to generation, accumulating in the bodies of living things in specific forms shaped by internal molds. The latter, determined by the form of the organism, dictated the way in which the parts had to be arranged in order to form a whole.

Consider for a moment the internal mold metaphor. Molds are used in smelting and masonry to impose a certain three-dimensional shape. Etymologically the word is related to modulus and model. In common usage it indicates a structural norm. In living organisms, however, the structural norm can accommodate irregularities, to which Buffon refers on more than one occasion as anomalies (*êtres anomaux*). An organic anomaly is not the same as a physical irregularity, however. Initially Buffon conceived of generation as analogous to crystallization, but ultimately he came to think of crystallization as a form of organization. He was unable to avoid associating anomalies with degeneration, hence with the problem of the mutability of species. On this point Buffon was never able to achieve certainty. He did not regard the idea of derivative species as absurd on its face, but he believed or professed to believe that observation confirmed the teachings of the Bible (see his article "The Ass" in the *Natural History of Animals*).

Maupertuis was bolder in theorizing, perhaps because he possessed less extensive empirical information. For him, structural variation was the rule of organic progression. In the *System of Nature* (1751) he set forth a theory of generation based on the existence of elementary particles of matter endowed with appetite and memory, whose "arrangement" reproduces the possibly miraculous structure of the first individuals (par. xxxi). The phenomena of resemblance, miscegenation, and monstrosity could be explained, he argued, in terms of the compatibility or incompatibility of "arrangements" in seeds mingled through copulation. "Can we not explain in this way how from just two individuals the most dissimilar species could have multiplied? Originally they may have stemmed from fortuitous productions in which the elementary parts did not retain the order they occupied in the father and mother animals. Each degree of error could have produced a new species, and repeated errors could have given rise to the infinite diversity of animals that we see today" (par. xlv).

It is tempting to read this text with spectacles provided by contemporary biochemical and genetic theory. *Order* and *error* occur both here and in contemporary accounts of hereditary biochemical defects as ground and cause of both normality and abnormality. But today biochemistry and genetics offer us a way of interpreting organic abnormalities that was worked out in cooperation with the Darwinian explanation of the origin of species and the adaptation of organisms. Hence Maupertuis's propositions should be regarded more as fictions than as anticipations of scientific theories to come. He was unable to overcome the difficulty posed by the natural mechanism for normalizing differences. Both he and Buffon believed that human intervention—through techniques of husbandry or agronomy—was the only way to stabilize variations within species: "What is certain is that any variety that might indicate a new species of animal or plant tends to die out. There are some deviations in which Nature persists only through art or government. Her own works always tend to regain the upper hand" (*Vénus physique*, 1745, part 2, chapter 5, conclusion). It was left to Darwin to discover variation, that is, a natural mechanism for normalizing minor anomalies.

The publication of *On the Origin of Species by Means of Natural Selection; or the Preservation of Favoured Races in the Struggle for Life* (1859) occasioned doubts in the minds of some early readers because of the traditional meaning of certain concepts mentioned in the title and frequently alluded to in the body of the work. The theory of natural selection states that certain deviations from the norm can be seen a posteriori to provide a tenuous advantage for survival in novel ecological situations. Darwin thus substituted a random fit for a preordained adaptation. Natural selection is eliminative. Disadvantaged organisms die; the survivors are all different in one degree or another. The reader who takes literally such Darwinian terms as *selection, advantage, adaptation, favor,* and *disfavor* may partially overlook the fact that teleology has been excluded from Darwin's theory. Does this mean that all value-laden terms have been excluded from the idea of life? Life and death, success or failure in the struggle for survival—are these value-neutral concepts, even if success is reduced to nothing more than continued existence? Does Darwin's language reveal his thought or does it suggest that even for Darwin a causal explanation of adaptation could not abolish the "vital meaning" of adaptation, a meaning determined by comparison of the living with the dead? As Darwin observed, variations in nature would have remained without effect, had it not been for natural selection. What could limit the ability of this law, operating over a long period of time and rigorously scrutinizing the structure, overall organization, and habits of every creature, to promote good and reject evil? (See *Origin of Species*, chapter 14, "Recapitulation," etc.)

And Darwin's work ends with a contrast, "that while this planet has gone circling on according to the fixed law of gravity, from so simple a beginning endless forms most beautiful and most wonderful have been and are being evolved."

In suggesting that individual variations, deviations in structure or instinct, are useful because they yield a survival advantage in a world in which relations of organism to organism are the most important of all causes of change in living beings, Darwin introduced a new criterion of normality into biology, a criterion based on the living creature's relation to life and death. By no means did he eliminate morality from consideration in determining the object of biology. Before Darwin, death was considered to be the regulator of the quantity of life (Buffon) or the sanction imposed for infractions of Nature's order, the instrument of her equilibrium (Linnaeus). According to Darwin, death is a blind sculptor of living forms, forms elaborated without preconceived idea, as deviations from normality are converted into chances for survival in a changed environment. Darwin purged from the concept of adaptation any reference to a preordained purpose, but he did not separate it completely from the concept of normality. In the spirit of Darwinism, however, a norm is not a fixed rule but a transitive capacity. The normality of a living thing is that quality of its relations to the environment that enables it to generate descendants exhibiting a range of variations and standing in a new relation to their respective environments, and so on. Normality is not a quality of the living thing itself but an aspect of the all-encompassing relation between life and death as it affects the individual life form at a given point in time.

Thus the environment decides, in a nonteleological way, which variations will survive, but this does not necessarily mean that evolution does not tend to create an organic order, firm in its orientation if precarious in its incarnations. Heredity is an uninterrupted delegation of ordinal power. What difference does it make if, in Salvador Luria's words, "evolution operates with threats, not promises."[5]

II

To sum up, in the 1860s both naturalists and physiologists looked upon the organism in two different ways, in terms of either normality or anomaly. But there was far from universal agreement as to the grounds for these divergent evaluations.

Because Darwinism refuted the claims of natural or revealed theology concerning the origin of species, some biologists and philosophers felt that it should also bear the burden of refuting any nonmechanist, nonmaterialist concept of life. In this they were encouraged by progress in the analysis and synthesis of organic compounds and in the reduction of vital processes to chemical transformations and exchanges of energy according to the emerging laws of thermodynamics.

The physiologists took their inspiration from a distinction first made by Bichat (*General Anatomy Applied to Physiology and Medicine*, 1801, vol. 1, pp. 20–21):

There are two kinds of life phenomena: (1) the state of health, and (2) the state of sickness. Hence there are two distinct sciences: physiology, which is concerned with phenomena of the first state, and pathology, which is concerned with those of the second. The history of phenomena in which the vital forces have their natural type leads us to that of phenomena in which those forces are distorted. Now, in the physical sciences, only the first history exists; the second is nowhere to be found. Physiology is to the motion of living bodies what astronomy, dynamics, hydraulics, hydrostatics, and so forth . . . are to the motions of inert bodies. The latter have no science that corresponds to them as pathology corresponds to the former.

But not all physiologists agreed with Bichat that there exist vital forces not subject to the laws of physics. Here I must cite Claude Bernard once more, because his position is so up-to-date. He admitted, first of all, that vital phenomena are subject only to physical and chemical causes, but he also held that the organism develops from the egg according to an immanent design, a plan, a regularity, which is responsible for its ultimate organization, for its harmony, persistence, and, if need be, restoration.

What Claude Bernard described in images is today explained by the theorems of macromolecular biochemistry. Like the metaphor of the "internal mold," the images of "design," "plan," "guiding idea," and "order" are given retroactive legitimacy by the concept of a program encoded in sequences of nucleotides.[6] For the first time in the history of biology, all the properties of living things—growth, organization, reproduction, hereditary continuity—can be explained in terms of molecular structure, chemical reactions, enzymes, and genes.

Yet twentieth-century biochemistry has reached a conclusion opposite to that toward which most nineteenth-century organic chemists were tending: the abolition of all differences between living and nonliving things. Today we recognize that living things exist in a state of unstable dynamic equilibrium, their orderly structure—whether of disorderly molecules or orderly crystals—maintained by means of a constant borrowing of energy from the environment. Thus, paradoxically, it was not simply because biology submitted its objects to the unfettered jurisdiction of physicists and chemists that the unique nature of those objects was established on a firm rational basis. The concepts of regulation and homeostasis were also required to make sense of the biological functions of resisting and delaying aging, disintegration, and disorder, relatively autonomous functions of open, hence environmentally dependent, living systems. Thus the intent behind all the intuitions, images, and metaphors of organic normality proved to be justified even as their content was shown to be of little value.

The level of objectivity at which the opposition between normal and abnormal was legitimate was shifted from the surface to the depths, from the developed organism to its germ, from the macroscopic to the ultramicroscopic. Now it is the transmission of the hereditary message, the production of the genetic program, that determines what is normal and what is a deviation from the normal. Some human chromosomal anomalies such as mongolism can be observed directly in the clinic. Others, such as Klinefelter's syndrome, are tolerated without apparent ill effect and manifest themselves only in special ecological circumstances. As for genetic anomalies, I shall mention only "innate errors of metabolism" (Garrod, 1909), that is, specific biochemical lesions that result from the presence of a mutant gene, which is called abnormal not so much because of its statistical rarity as because of its pathological or even fatal effects (hemophilia, Huntington's chorea, etc.). A new nomenclature of disease is thus established, referring disease not to the individual considered in its totality but to its morphological and functional constituents: diseases of the hemoglobin, hormonal diseases (such as hyperthyroidism), muscle diseases, and so on. Gene mutations that block chemical syntheses by altering their enzyme catalysts are no longer interpreted as deviations in Maupertuis's sense but as errors in reading the genetic "message," errors in the reproduction or copying of a text.

The term *error* does not imply that science has returned to the Aristotelian and medieval notion that monsters are errors of nature, for the failure here is not some lack of skill on the part of the artisan or architect but a mere copyist's slip. Still, the new science of living things has not only not eliminated the contrast between normal and abnormal; it has actually grounded that contrast in the structure of living things themselves.[7]

III

A remarkable and interesting fact from the epistemological standpoint is the proliferation of terms containing the prefix *auto-*, used today by biologists to describe the functions and behavior of organized systems: auto-organization, auto-reproduction, auto-regulation, auto-immunization, and so on. No doubt biophysicists and biochemists have been searching for terms to express the mechanisms underlying these properties and to construct cybernetic models of self-reproductive automata (J. von Neumann). These models are only logical, however, and the only actual self-reproductive automata are natural organic systems, that is, living organisms. The epistemological reason for preceding these terms with the prefix *auto-* is to convey something about the nature of their relation to the environment. According to Schrödinger, "Life is a behavior of matter . . . based upon the maintenance of a preexisting order" (*What Is Life?*, 1945); and according to A. Lwoff, "The biological order has no source other than the biological order" (*L'ordre biologique*, 1962). Living systems are open, non-equilibrium systems that maintain their organization both *because* they are open to the external world and *in spite of* being open to the external world.[8] Organization by whatever name one wishes to call it—negentropy, information, systemic improbability—expresses the quality of a certain physical quantity. That alone suffices to distinguish biology from physics, even though the former now seems to have linked its destiny to the latter. The biologist cannot help continuing to use the concept of normality. Suppose, for example, that one base in the genetic sequence is substituted for another. Lwoff points out that "for the physicist, even if the mutation is lethal, nothing has changed. The quantity of negative entropy has not varied. But since the mutation is lethal, the transformed organism cannot function normally or reproduce itself. It has ceased to live" (ibid.). Or think of Léon Brillouin's related example of a skillful surgeon who is able to separate the organs of an animal, keep them alive, and then reassemble them to create either a viable being or a monstrous creature that cannot sustain life: "The two reconstructions are equally improbable, but the value of the first is higher than that of the second. Should the definition of total negentropy be associated with improbability or value? Shall we consider a monster the equivalent of a 'well-balanced' being? Only the notion of value seems to fit this new problem, but how are to define it properly?"[9]

IV

Perhaps the epistemologist may now be allowed to remain skeptical about dogmatic reductionist views, given what can be learned if we look at the history of biology without any simplifying a priori assumptions and in light of the various manifestations of what I have proposed calling the principle of thematic conservation.

I anticipate one possible objection, however. In looking for a distinctive concept of normality in biology, have I not confused the issue by considering different orders of biological objects? Astronomers from Herschel to Hubble revolutionized their discipline by magnifying their object to an unimaginable degree, revealing galaxies beyond the solar system and metagalaxies beyond the galaxies. By contrast, biologists have discovered the nature of life by making their objects smaller and smaller: bacterium, gene, enzyme. In the preceding discussion, am I dealing with observations at one level and explanations at another? Normality appears to be a property of an organism, but it disappears when we look at the elements that make up that organism.

At all levels, however, biologists have identified ordering structures that while generally reliable sometimes fail. The concept of normality is intended to refer to these ordering structures. No such concept is needed in the epistemology of physics. By introducing it as I have done here I in no way intend to deny that biology is based on physics and chemistry. I do intend to prevent the coalescing of two properly distinct approaches to history. In the history of biology the pseudotheoretical content of prescientific conceptualizations of structural and functional normality was abandoned, but the conceptualizations themselves have been preserved, in "displaced" form, as indices of the objective uniqueness of the living organism. Mendeleev's periodic table does not justify Democritus's intuitions a posteriori, but the decoding of the genetic program does provide a posteriori justification of Claude Bernard's metaphors. Even within the terms of a monist, indeed a materialist, epistemology, physics remains radically different from biology. Physics was produced, sometimes at risk of life and limb, by living things subject to sickness and death, but sickness and death are not problems of physics. They are problems of biology.

Between the bacteria in a laboratory culture and the biologists who observe them there is a whole range of living things permitted to exist by the filter of natural selection. Their lives are governed by certain norms of behavior and adaptation. Questions about the vital meaning of those norms, though not directly matters of chemistry and physics, are questions of biology. As Marjorie Greene points out, alongside the biochemists there is room in biology for a Buytendijk or a Kurt Goldstein.[10] History shows that she is right.

My purpose in this essay was in part to show how philosophy can influence the statement of a historical problem, in this case a question in the history of biology. It may be that I failed to achieve this goal. But I also wanted to challenge the view that there is no point in asking such questions, which only complicate matters needlessly. For I maintain that the proper function of philosophy is precisely to complicate matters, not only for the historian of science but for man in general.

The Living and Its Milieu

Georges Canguilhem

Georges Canguilhem (1904–1995) was a preeminent philosopher and
historian of science and medicine. Trained as a doctor as well as a
philosopher, he developed an original manner of inquiring into problems
of biology, physiology and medicine (among other topics) through the
historical study of theories, concepts, and objects. In this essay, Canguilhem
takes up the problematic relation of knowledge and life, first through a
history of the concept of "milieu" and then through the identification of the
maladaptation and "pathos" in the living being's relation to the milieus in
which it finds itself. Scientific knowledge, he concludes, must be understood
as the project of a living, sensing being, a being striving to constitute its
own vital environment.

The notion of milieu[1] is becoming a universal and obligatory mode of appre-
hending the experience and existence of living beings; one could almost say
it is now being constituted as a category of contemporary thought. But until
now it has been quite difficult to perceive in a synthetic unity the historical
stages of the formation of this concept, the various forms of its utilization,
and the successive inversions of the relationship in which it is one of the
terms—in geography, in biology, in psychology, in technology, in economic
and social history. For this reason, philosophy must take the initiative in
synoptically investigating the meaning and value of this concept. By initiative, we do not mean what appears to be an initiative but only consists in
reflecting on the sequence of scientific explorations so as to compare their
appearance and results. Rather, through a critical comparison of several ap-

proaches, we mean, if possible, to bring to light their common point of departure, and to postulate their fecundity for a philosophy of nature centered on the problem of individuality. We shall thus examine one by one the simultaneous and successive components of the notion of *milieu*, the varieties of its use, from 1800 to our time, the various reversals of the relationship between organism and milieu, and, finally, the general philosophical impact of these reversals.

Historically speaking, the notion and term *milieu* were imported from mechanics into biology during the second half of the eighteenth century. The mechanical notion (though not the term) appeared with Newton, and in its mechanical meaning the term can be found in the article "Milieu" in d'Alembert and Diderot's *Encyclopédie*.[2] Lamarck, inspired by Buffon, introduced it into biology, but he used it only in the plural. This usage was established by Henri de Blainville. Étienne Geoffroy Saint-Hilaire (in 1831) and Auguste Comte (in 1838) used the term in the singular, as an abstract term. Honoré de Balzac introduced it into literature in 1842 (in the preface to *The Human Comedy*), and Hippolyte Taine established it as one of the three principles of the analytic explanation of history—the other two being race and moment.[3] It is from Taine, rather than from Lamarck, that French neo-Lamarckian biologists after 1870—Alfred Giard, Félix Le Dantec, Frédéric Houssay, Johann Costantin, Gaston Bonnier, Louis Roule—inherited this term. The idea came from Lamarck, but the term, as universal and abstract, was transmitted to them by Taine.

The French mechanists of the eighteenth century called "milieu" what Newton had referred to as "fluid." In Newton's physics, the type—if not the sole archetype—of fluid is ether.[4] In Newton's time, the problem mechanics had to solve was that of the action of distinct physical bodies at a distance. This was the fundamental problem in the physics of central forces. It was not an issue for Descartes, however. For him, there is but one mode of physical action, collision, in one possible physical situation, contact. This is why we can say that the notion of milieu has no place in Cartesian physics. Descartes' "subtle matter" is in no way a milieu. But there was difficulty in extending the Cartesian theory of collision and contact to the case of distinct physical bodies, for their actions blend together. We thus understand how Newton came to pose the problem of the medium of action.[5] For him, luminiferous ether is fluid as the medium of action at a distance. This explains the passage from the notion of fluid as vehicle to that of its designation as milieu. The fluid is an intermediary between two bodies; it is their milieu; and insofar as the fluid penetrates all these bodies, they are situated in the middle of it [*au milieu de lui*]. According to Newton and the physics of central forces, one can speak of an environment, a milieu, because there exist centers of force. The notion of milieu is an essentially relative one. When we consider separately the body which receives an action transmitted by the milieu, we forget that a *milieu* is a medium, *in between two centers*, and we retain only its function as a centripetal transmitter, its position as that which surrounds a body. In this way, milieu tends to lose its relative meaning and to take on that of an absolute, a reality in itself.

Newton is perhaps responsible for the importation of the term from physics into biology. He used ether not only to solve the problem of the phenomenon of illumination but also to explain the physiological phenomenon of vision and, finally, to explain the physiological effects of the sensation of light, that is, muscular reactions. In his *Optics*, Newton considers ether to be continuous in the air, the eye, the nerves, and the muscles. It is thus the action of the milieu that guarantees the relation of dependence between the illumination of a perceived light source and the movement of the muscles by which man reacts to this sensation. This, it seems, is the first example of an organic reaction being explained by the action of a milieu, that is to say, by the action of a fluid strictly defined by physical properties.[6] Indeed, the aforementioned *Encyclopédie* article confirms this view and borrows all its examples of a milieu from Newton's physics. And it is in a purely mechanical sense that water is said to be a milieu for the fish that move about in it. It is also in this mechanical sense that Lamarck first uses the term.

Lamarck always speaks of milieus—in the plural—by which he expressly means fluids like water, air, and light. When Lamarck wishes to designate the ensemble of actions that act on a living being from the outside—what we today call the milieu—he never says "milieu," but always "influencing circumstances."[7] Consequently, circumstance is for Lamarck a genus, whose species are climate, place, and milieu. This is why Léon Brunschvicg, in *Les étapes de la philosophie mathématique*[8] could write that Lamarck had borrowed from Newton the model for a physical-mathematical explanation of the living by a system of connections with its environment. The connections between Lamarck and Newton are direct at the intellectual level and indirect historically. Buffon links Lamarck to Newton. We might simply recall that Lamarck was Buffon's student and the tutor of his son.

Buffon in fact combines two influences in his conception of the relations between the organism and the milieu. The first is Newton's cosmology, of which Buffon was a constant admirer. The second is the tradition of anthropo-geographers, which, after Machiavelli, Jean Bodin, and John Arbuthnot, was kept alive in France by Montesquieu. The Hippocratic treatise *On Airs, Waters, and Places* can be considered the first work to have given a philosophical form to this anthropo-geographic conception.[9] These are the two elements Buffon brought together in his principles of animal ethology, to the extent that the mores of animals are distinctive and specific characteristics and can be explained by the same method geographers use to explain the diversity of men—the variety of races and peoples on the earth's surface.[10]

Thus, as teacher and precursor to Lamarck in his theory of milieu, Buffon appears at the convergence of the two components of this theory: the mechanical and the anthropo-geographical. Here is posed a problem of epistemology and of the historical psychology of knowledge, a problem whose scope greatly exceeds the present example. Shouldn't we interpret the fact that two or more guiding ideas combine at a certain moment to form a single theory as a sign that—in the final analysis and despite their apparent differences—they have a common origin, whose meaning and very existence we forget when we consider separately their disjointed parts? We will return to this problem at the end of the present essay.

The Newtonian origins of the notion of milieu thus suffice to account for its initial mechanical signification and the use that was first made of it. The origin determines the meaning and the meaning determines the usage, to such an extent that Comte, when proposing a general biological theory of milieu in 1838 (in the 40th lesson of his *Course of Positive Philosophy*), had the impression he was using *milieu* as a neologism and claimed responsibility for erecting it into a universal and abstract notion of biological explanation. Comte says that by this term he no longer means only "the fluid into which a body is immersed" (thereby confirming the mechanical origins of the notion) but "the total ensemble of exterior circumstances necessary for the existence of each organism." But we also see in Comte—who has a perfectly clear sense of the origins of the notion, as well as of the import he would like to give to it in biology—that its usage will remain dominated by the mechanical origins of the notion, if not of the term. Indeed, it is quite interesting to notice that Comte is on the brink of forming a dialectical conception of the relations between the organism and the milieu. We are alluding here to the passages in which he defines the relation of the "appropriate organism" and the "suitable milieu" as a "conflict of forces," and the act constituting that conflict as function.[11] He posits that "the ambient system could not possibly modify the organism if the organism did not exert on it in turn a corresponding influence." But, apart from the case of the human species, he holds the organism's action on the milieu to be negligible. In the case of the human species, Comte, faithful to his philosophical conception of history, admits that, by the intermediary of collective action, humanity modifies its milieu. Still, for the living in general Comte refuses to consider this reaction of the organism on the milieu—judging it to be simply negligible. This is because he very explicitly looks for a guarantee of this dialectical link, this reciprocal relation between milieu and organism, in the Newtonian principle of action and reaction. Indeed from a mechanical point of view, the action of the living on the milieu is almost negligible. And Comte ends up posing the biological problem of the relations between the organism and the milieu in the form of a mathematical problem: "In a given milieu, and given an organ, find the function—and vice versa." The link between the organism and the milieu is thus that of a function to an ensemble of variables, an equation by way of which, "all other things being equal," one can determine the function by the variables, and each variable by the function.[12]

In the forty-third lesson of the *Course of Positive Philosophy*, Comte analyses the variables for which the milieu is the function. These variables are weight, air and water pressure, movement, heat, electricity, and chemical species—all factors that can be studied experimentally and quantified by measurements. The quality of an organism is reduced to an ensemble of quantities, despite Comte's professed distrust of the mathematical treatment of biological problems—a distrust that came to him from Bichat.

In sum, the benefit of even a cursory history of the importation of the term *milieu* into biology during the first years of the nineteenth century is that it accounts for the originally strictly mechanistic acceptance of the term. If in Comte there appears a hint of an authentically biological acceptance and a more flexible usage of the word, this immediately gives way to the prestige of mechanics, an exact science in which prediction is based on calculation. To Comte, the theory of milieu seems clearly to be a variant of the fundamental project that the *Course of Positive Philosophy* endeavors to complete: first the world, then man; to go from the world to man. If Comte anticipates the idea of a subordination of the mechanical to the vital—the idea he would later formulate in mythical form in *The System of Positive Polity* and *The Subjective Synthesis*—here he nevertheless deliberately represses it.

But there is still one lesson to be taken from the use—absolute and without qualification—of the term *milieu* as it was definitively established by Comte. The term would henceforth designate the equivalent of Lamarck's "circumstances" and Étienne Geoffroy Saint-Hilaire's "ambient milieu" (in his 1831 thesis at the Académie des Sciences). These terms, *circumstances* and *ambience*, point to a certain intuition of a formation around a center. With the success of the term *milieu*, the representation of an indefinitely extendible line or plane, at once continuous and homogeneous, and with neither definite shape nor privileged position, prevailed over the representation of a sphere or circle, which are qualitatively defined forms, and, dare we say, attached to a fixed center of reference. *Circumstances* and *ambience* still retain a symbolic value, but *milieu* does not evoke any relation except that of a position endlessly negated by exteriority. The now refers to the before; the here refers to its beyond, and thus always and ceaselessly. The milieu is truly a pure system of relations without supports.

From there one can understand the prestige of the notion of milieu for analytic scientific thought. The milieu becomes a universal instrument for the dissolution of individualized organic syntheses into the anonymity of universal elements and movements. When the French neo-Lamarckians borrowed from Lamarck, if not the term *milieu* in the singular and in its absolute sense, then at least the idea of it, they retained of the morphological characteristics and functions of the living only their formation by exterior conditioning—only, so to speak, their formation by deformation. It is enough to recall J. Costantin's experiments on the forms of the arrowhead leaf or Frédéric Houssay's experiments on the form, fins, and metamerism of fish.[13] Louis Roule was able to write, in his small book *La vie des rivières*,[14] that "fish do not lead their lives on their own, it is the river that makes them lead it; they are persons without personality." We have here an example of what a strictly mechanist usage of the notion of milieu necessarily leads to.[15] We are brought back to the theory of animal-machines. In the end, this is just what Descartes said, when he said of animals that "it is nature which acts in them by means of their organs."[16]

From 1859 on—that is to say, after the publication of Darwin's *The Origin of Species*—the problem of the relations between organism and milieu is dominated by the polemic between Lamarckians and Darwinians. To understand the meaning and importance of this polemic, it is necessary to recall the originality of their respective points of departure.

In his 1809 *Zoological Philosophy*, Lamarck writes that if by action of circumstances or milieus, one takes him to mean direct action by the exterior milieu on the living, one is putting words into his mouth.[17] It is via the intermediary of need, a subjective notion implying reference to a positive pole of vital values, that the milieu dominates and compels the evolution of living beings. Changes in circumstances lead to changes in needs; changes in needs lead to changes in actions. If these actions are long-lasting, the use or nonuse of certain organs causes the organs to develop or atrophy, and these morphological acquisitions or losses, obtained by individual habit, are preserved by the mechanism of heredity, on condition that the new morphological characteristic is common to both parents.

According to Lamarck, the situation of the living in the milieu is distressful and distressed. Life exists in a milieu that ignores it, as two asynchronous series of events. Circumstances change on their own, and the living must take the initiative not to be "dropped" by its milieu. Adaptation is a renewed effort by life to continue to "stick" to an indifferent milieu. Since it is the result of an effort, adaptation is thus neither harmonious nor providential; it is gained and never guaranteed. Lamarckism is not mechanist, and it would also be inaccurate to call it finalistic. In reality, it is a bare vitalism. There is an originality in life for which the milieu does not account and which it ignores. Here the milieu is truly exterior, in the proper sense of the word: it is foreign, it does nothing for life. This is truly a vitalism because it is a dualism. Life, says Bichat, is the ensemble of functions that resist death. In Lamarck's conception, life resists solely by deforming itself so as to outlive itself. To our knowledge, no portrait of Lamarck, no summary of his doctrine, surpasses the one given by Charles Augustin Sainte-Beuve in his novel *Volupté*.[18] One sees how far one has to go to get from Lamarck's vitalism to the French neo-Lamarckians' mechanism. Edward Cope, an American neo-Lamarckian, was more faithful to the spirit of the doctrine.

Darwin had a completely different idea of the environment of the living, as well as of the appearance of new forms. In the introduction to *The Origin of Species*, he writes: "Naturalists continually refer to external conditions such as climate, food, etc. as the only possible cause of variation. In one limited sense, . . . this may be true."[19] It seems that Darwin later regretted having attributed only a secondary role to the direct action of physical forces on the living. This comes across in his correspondence. Marcel Prenant, in his introduction to a collection of Darwin's texts, has published some particularly interesting passages on this topic.[20] Darwin looks for the appearance of new forms in the conjunction of two mechanisms: one that produces differences, namely, variation; and one that reduces and tests the differences thereby produced, namely, the struggle for life and natural selection. The fundamental biological relation, in Darwin's eyes, is the relation of one living being to others; it prevails over the relation between the living and the milieu conceived as an ensemble of physical forces. The first milieu an organism lives in is an entourage of living beings, which are for it enemies or allies, prey or predators. Between these living beings are established relations of use, destruction, and defense. In this competition of forces, accidental morphological variations count as advantages or disadvantages. And variation—the appearance of small morphological differences by which a descendant does not exactly resemble its ancestors—stems from a complex mechanism: the use or nonuse of organs (the Lamarckian factor only applies to adults), correlations or compensations in growth (for the young), or the direct action of the milieu (on germ seeds).

In this sense, one can say that for Darwin, by contrast to Lamarck, the initiative to variation comes sometimes—but only sometimes—from the milieu. One gets a somewhat different idea of Darwin depending on whether one accentuates this action or not and whether one limits oneself to his classic works or instead considers the entirety of his thought, as revealed in his correspondence. In any case, for Darwin, to live is to submit an individual difference to the judgment of the ensemble of living beings. This judgment has only two possible outcomes: either death or becoming oneself part of the jury for a while. So long as one lives, one is always judge and judged. As a result, in Darwin's oeuvre as he left it to us, the thread linking the formation of the living being to the physico-chemical milieu can seem fairly thin. And when mutationism, a new theory of the evolution of species, used genetics to explain the appearance of immediately hereditary species variations (Darwin had underestimated this phenomenon), the role of the milieu was reduced to eliminating the worst without participating in the production of new beings, normalized by their unpremeditated adaptation to new conditions of existence, monstrosity becoming the rule and originality a temporary banality.

In the polemic between Lamarckians and Darwinians, the same arguments and objections are made in both directions and are applied to both authors: finalism is denounced and mechanism celebrated sometimes in one, sometimes in the other. This is no doubt a sign that the question has been badly put. In Darwin, one can find finalism not in things themselves but in his choice of words—he has been frequently reproached for his term *selection*. In Lamarck, it is less finalism than vitalism. Both are authentic biologists, to whom life appears as a given that each seeks to characterize, instead of trying analytically to explain it. These two authentic biologists are complementary. Lamarck thinks of life in terms of duration, and Darwin thinks of it mostly in terms of interdependence: a living form presupposes a plurality of other forms in relation to it. The synoptic vision that is the essence of Darwin's genius is missing in Lamarck. Darwin is more closely related to the geographers, and we know how much he owed to his voyages and explorations. The milieu in which Darwin depicts the life of the living is a bio-geographical milieu.

At the beginning of the nineteenth century, two names stand for the birth of geography as a science conscious of its method and dignity: Carl Ritter and Alexander von Humboldt.

In 1817, Ritter published his *Comparative Geography*.[21] Humboldt published, during the decade beginning in 1845, a book whose title, *Kosmos*, perfectly captures its spirit.[22] In these two works are united the traditions of Greek geography: that is to say, on the one hand, the science of the human ecumene since Aristotle and Strabo, and on the other, the science of the co-ordination of human space in relation to celestial configurations and movements—the science of mathematical geography, which Eratosthenes, Hipparchus, and Ptolemy are considered to have founded.

According to Ritter, without man's relation to the land—to all land—human history is unintelligible. The earth, considered as a whole, is the stable ground for the vicissitudes of history. Terrestrial space and its configuration are, consequently, not only geometrical and geological objects of knowledge but also sociological and biological ones.

Humboldt was a naturalist-traveler who repeatedly covered what was possible to cover of the world in his time and who applied a whole system of barometric, thermometric, and other measurements to his investigations. Humboldt's interest was above all focused on the distribution of plants according to climate: he is the founder of botanical geography and zoological geography. *Kosmos* is a synthesis of knowledge concerning life on earth and the relations of life to the physical milieu. This synthesis does not aim to be an encyclopedia but rather to arrive at an intuition of the universe; it begins with a history of *Weltanschauungen*, with a history of the Cosmos whose equivalent it would be difficult to find in a work of philosophy. It is an absolutely remarkable overview.

It is essential to note that Ritter and Humboldt applied to their object—the relations between historical man and milieu—the category of totality. Their object is the whole of humanity on the whole Earth. With Ritter and Humboldt, the idea of determining historical relations by the geographical substrate was consolidated in geography. It gave rise first to Friedrich Ratzel and anthropo-geography in Germany, and then to geopolitics. The idea then invaded history by contagion, starting with Michelet (let us recall his *Le tableau de la France*).[23] And finally, as we have already said, Taine contributed to the spread of the idea to all milieus, including the literary milieu. We can sum up the spirit of this theory of the relations of geographical milieu to man by saying that doing history came to consist in reading a map, where this map is the figuration of an ensemble of metrical, geodesic, geological, and climatological data, as well as descriptive bio-geographical data.

The treatment of anthropological and human ethological questions—a treatment that became more and more deterministic, or rather, mechanistic the farther one went from the spirit of its founders—was doubled by a parallel, if not exactly synchronous treatment in the domain of animal ethology. The mechanistic explanation of the organism's movements in the milieu succeeded the mechanistic interpretation of the formation of organic forms. Let us simply recall the works of Jacques Loeb and John B. Watson. Generalizing the conclusions of his research on phototropisms in animals, Loeb considered all movement of the organism as movement forced on it by the milieu. The reflex, considered as an elementary response of a segment of the body to an elementary physical stimulus, is the simple mechanism whose composition allows one to explain all behaviors of the living. Along with Darwinism, this exorbitant Cartesianism is incontestably at the origin of the postulates of behaviorist psychology.[24]

Watson assigned psychology the task of conducting analytic research into the conditions of the adaptation of the living to the milieu by experimentally producing excitation and response relations (the stimulus-response pair). There is a physical determinism in the relation between excitation and response. The biology of behavior is reduced to neurology, which itself is reducible to energetics, the science of energy. The evolution of Watson's thought led him from a conception which simply neglects consciousness as useless to one that nullifies it as illusory. The milieu thus comes to be invested with all power over individuals; its power [*puissance*] dominates and even abolishes that of heredity and genetic constitution. Since the milieu is given, the organism gives itself nothing it does not, in reality, already receive. The situation of the living, its being in the world, is a condition, or, more exactly, a conditioning.

Albert Weiss intended to build biology like a deductive physics, by proposing an electronic theory of behavior. It fell to the psycho-technicians—who expanded Taylorist techniques for timing movements through the analytic study of human reactions—to perfect the work of behaviorist psychology, and constitute, through their science, man as a machine reacting to machines, as an organism determined by the "new milieu" (Friedmann).

In short, because of its origins, the notion of milieu first developed and spread in a perfectly determined way, and we can say, applying to this notion the methodological norm it stands for, that its intellectual power was a function of the intellectual milieu in which it had been formed. The theory of milieu was at first the positive and apparently verifiable translation of Condillac's fable of the statue.[25] When the air smells like roses, a statue is rose-scented. In the same way, the living, within the physical milieu, is light and heat, carbon and oxygen, calcium and weight. It responds by muscular contractions to sensory excitations; it responds with a scratch to an itch, with flight to an explosion. But one can and must ask: Where is the living? We see individuals, but these are objects; we see gestures, but these are displacements; centers, but these are environments; machinists, but these are machines. The milieu of behavior coincides with the geographical milieu; the geographical milieu, with the physical milieu.

It was normal, in the strong sense of the word, for this methodological norm to have first reached its limits and the occasion for its reversal in geography. Geography has to do with complexes—complexes of elements whose actions mutually limit each other and in which the effects of causes become causes in turn, modifying the causes that gave rise to them. Trade winds are a typical example of a complex in this respect. They displace surface water that has been heated by contact with the air; the cold deep waters rise to the surface and cool the atmosphere; low temperatures engender low pressure, which generates winds; the cycle is closed and begins again. The same type of complex can be observed in plant geography. Vegetation grows in natural ensembles, in which different species limit each other reciprocally and where, in consequence, each contributes to creating an equilibrium for the others. The ensemble of these plant species ends up constituting its own milieu. Thus the exchanges between plants and the atmosphere end up creating a sort of vapor screen around the vegetal zone, which limits the effect of radiation, and this cause gives rise to an effect that will in turn slow down the cause, and so on.[26]

The same approaches must be applied to animals and to man. However, the human reaction to provocation by the milieu is diversified. Man can give several different solutions to a single problem posed by the milieu. The milieu proposes, without ever imposing, a solution. To be sure, in a given state of civilization and culture, the possibilities are not unlimited. But the fact of considering as an obstacle something that may later be seen as a means to action ultimately derives from the idea, the representation, that man (collective man, of course) builds himself out of his possibilities, his needs. In short, it results from what he represents to himself as desirable, which is inseparable from the ensemble of values.[27]

Thus, the relation between the milieu and the living being ends up reversed. Man, as a historical being, becomes the creator of a geographical configuration; he becomes a geographical factor. We simply call to mind here that the works of Paul Vidal-Lablache, Jean Brunhes, Albert Demangeon, and Lucien Febvre and his school have shown that, for man, there is no pure physical milieu. Within a human milieu, man is obviously subjected to a kind of determinism, but this is the determinism of artificial creations, from which the spirit of invention that brought them into existence has been alienated. In this same line of thought, the work of Friedmann shows how, in the new milieu that machines create for man, the same reversal has already been brought about. Pushed to the extreme limits of its ambition, the engineers' psycho-technics that descended from Taylor's ideas succeeds in grasping, as an irreducible center of resistance, the presence in man of man's own originality in the form of a sense of values. Even when subordinated to machines, man cannot apprehend himself as a machine. His productive efficiency improves the better aware he is of his centrality with regard to mechanisms intended to serve him.

Much earlier, the same reversal of the relation between organism and milieu had taken place in animal psychology and the study of behavior. Jacques Loeb led to Herbert Spencer Jennings, and John B. Watson to Robert Jacob Kantor and Edward C. Tolman.

Here, the influence of pragmatism is obvious and well established. If pragmatism served as an intermediary between Darwinism and behaviorism—in one sense by generalizing and extending the notion of adaptation to the theory of knowledge, and in another by emphasizing the role of values in relation to the interests of an action—John Dewey led the behaviorists to see the reference of organic movements to the organism itself as essential. The organism is considered a being on which not everything can be imposed, because its existence as organism consists in its proposing itself to things on the basis of certain orientations that are proper to it. Tolman's teleological behaviorism, first developed by Kantor, consists in searching for and recognizing the meaning and intention of animal movement. What appears essential in the movement of reaction is that it persists, through a variety of phases, which can be errors or lapses, until the moment when the reaction either brings the excitation to an end and re-establishes rest or leads to a new series of acts, entirely different from those that have been concluded.

Before Tolman, Jennings, in his theory of trial and error, had shown (against Loeb) that the animal does not react as a sum of distinct molecular reactions to a stimulant that can be divided into units of excitation. Instead, the animal reacts as a whole to total objects, and its reactions are regulators for the needs that govern them. Naturally, one must recognize here the considerable contribution of *Gestalttheorie*, and in particular of Kurt Koffka's distinction between the milieu of behavior and the geographical milieu.[28]

Finally, the relation between organism and milieu is reversed in von Uexküll's studies of animal psychology and in Goldstein's studies of human pathology. Each of them makes this reversal with a lucidity that comes from a fully philosophical view of the problem. Von Uexküll and Goldstein agree on this fundamental point: to study a living being in experimentally constructed conditions is to make a milieu for it, to impose a milieu on it; yet it is characteristic of the living that it makes its milieu for itself, that it composes its milieu. Of course, we might still speak of interaction between the living and the milieu even from a materialist point of view—between one physico-chemical system cut out from a larger whole, and its environment. But to speak of interaction does not suffice to annul the difference between a relation of the physical type and a relation of the biological type.

From the biological point of view, one must understand that the relationship between the organism and the environment is the same as that between the parts and the whole of an organism. The individuality of the living does not stop at its ectodermic borders any more than it begins at the cell. The biological relationship between the being and its milieu is a functional relationship, and thereby a mobile one; its terms successively exchange roles. The cell is a milieu for intracellular elements; it itself lives in an interior milieu, which is sometimes on the scale of the organ and sometimes of the organism; the organism itself lives in a milieu that, in a certain fashion, is to the organism what the organism is to its components. In order to judge biological problems, we thus require a biological sense, to whose formation von Uexküll and Goldstein can greatly contribute.[29]

Von Uexküll chooses the words *Umwelt*, *Umgebung*, and *Welt* and distinguishes between them with great care. *Umwelt* designates the milieu of behavior proper to a certain organism; *Umgebung* is the banal geographical environment; *Welt* is the universe of science. The milieu of behavior proper to the living (*Umwelt*) is an ensemble of excitations, which have the value and signification of signals. To act on a living being, a physical excitation has not only to occur but also to be noticed. Consequently, insofar as the excitation acts on the living being, it presupposes the orientation of the living being's interest; the excitation comes not from the object, but from the living. In order for the excitation to be effective, it must be anticipated by an attitude of the subject. If the living is not looking, it will not receive anything. A living being is not a machine that responds to excitations with movements, it is a machinist, who responds to signals with operations. Naturally, this is not to contest that it happens through reflexes whose mechanism is physico-chemical. That is not where the question lies for the biologist. Rather, the question lies in the fact that out of the abundance of the physical milieu, which produces a theoretically unlimited number of excitations, the animal retains only some signals (*Merkmale*). Its life rhythm orders the time of this *Umwelt*, just as it orders space. Along with Buffon, Lamarck used to say that time and favorable circumstances constitute the living bit by bit. Von Uexküll turns the relation around and says: time and favorable circumstances are relative to certain living beings.

The *Umwelt* is thus an elective extraction from the *Umgebung*, the geographical environment. But the environment is precisely nothing other than the *Umwelt* of man, that is to say, the ordinary world of his perspective and pragmatic experience. Just as this *Umgebung*, this geographic environment external to the animal, is, in a sense, centered, ordered, oriented by a human subject—that is to say, a creator of techniques and a creator of values—the *Umwelt* of the animal is nothing other than a milieu centered in relation to that subject of vital values in which the living essentially consists. We must see at the root of this organization of the animal *Umwelt* a subjectivity analogous to the one we are bound to see at the root of the human *Umwelt*. One of the most gripping examples cited by von Uexküll is the *Umwelt* of the tick.

Ticks live off the warm blood of mammals. The adult female, after mating, climbs to the end of a tree branch and waits. She can wait up to eighteen years. At the Rostock Institute of Zoology, ticks were kept alive in captivity without eating for eighteen years. When a mammal passes under the tick's lookout and hunting post, she drops down. It is the smell of rancid butter emanating from the animal's coetaneous glands that guides her. This is the only stimulant which can set off this falling movement. This is the first stage. When she has fallen onto the animal, she attaches herself there. If the odor of rancid butter has been artificially produced—on a table, for example—the tick will not stay there, but will climb back up to her observation post. Only the temperature of the blood keeps her on the animal. She is fixed to the animal by her thermal sense and, guided by her tactile sense, she seeks out places on the skin where there are no hairs. She buries her head there, and sucks the blood. Only at the moment when the mammal's blood enters into her stomach do the tick eggs (encapsulated ever since the moment of mating and capable of remaining encapsulated for eighteen years) open, mature, and develop. The tick can live for eighteen years to perform her reproductive function in several hours. It is noteworthy that, over a long period of time, the animal can remain totally indifferent, insensible to all the excitations that emanate from a milieu such as the forest, and that the sole excitation which can release its movement—to the exclusion of all others—is the odor of rancid butter.[30]

A comparison with Goldstein is imperative here, for Goldstein bases his theory on a critique of the mechanical theory of reflexes. A reflex is not an isolated or gratuitous reaction. A reaction is always a function of the opening of a sense to stimulations, and of its orientation with regard to them. This orientation depends on the signification of a situation indistinct from this ensemble. Isolated stimuli have meaning for human science, but none for the sensibility of a living being. An animal in an experimental situation is in an abnormal situation, a situation it does not need according to its own norms; it has not chosen this situation, which is imposed on it. An organism is thus never equal to the theoretical totality of its possibilities. One cannot understand its actions without appealing to the notion of privileged behavior. "Privileged" does not mean objectively simpler—just the inverse. The animal finds it simpler to do what it privileges. It has its own vital norms.

The relation between the living and the milieu establishes itself as a debate (*Auseinandersetzung*), to which the living brings its own proper norms of appreciating situations, both dominating the milieu and accommodating itself to it. This relation does not essentially consist (as one might think) in a struggle, in an opposition. That applies to the pathological state. A life that affirms itself against the milieu is a life already threatened. Movements of force—for example, reactions of muscular extension—translate the exterior's domination of the organism.[31] A healthy life, a life confident in its existence, in its values, is a life of flexion, suppleness, almost softness. The situation of a living being commanded from the outside by the milieu is what Goldstein considers the archetype of a catastrophic situation. And that is the situation of the living in a laboratory. The relations between the living and the milieu as they are studied experimentally, objectively, are, among all possible relations, those that make the least sense biologically; they are pathological relations. Goldstein says that, in the organism, "'meaning' and 'being' are the same"; we can say that the being of an organism is its meaning.[32] Certainly, the living can and must be analyzed in physico-chemical terms. This has its theoretical and practical interest. But this analysis is a chapter in physics. In biology, everything is still to be done. Biology must first hold the living to be a significative being, and it must treat individuality not as an object but as an attribute within the order of values. To live is to radiate; it is to organize the milieu from and around a center of reference, which cannot itself be referred to without losing its original meaning.

While the relation between organism and milieu was being reversed in animal ethology and in the study of behavior, the explanation of morphological characteristics was undergoing a revolution that led to the acceptance of the autonomy of the living in relation to the milieu. We are alluding here to the well-known works of William Bateson, Lucien Cuénot, Thomas Hunt Morgan, Hermann Müller, and their collaborators, who took up and extended Gregor Mendel's research on hybridization and heredity.[33] In creating the science of genetics, they came to maintain that the acquisition by the living being of its form and, hence, its function depends, in a given milieu, on its own hereditary potential and that the milieu's action on the phenotype leaves the genotype intact. The genetic explanation of heredity and evolution (the theory of mutations) converged with August Weismann's theory. Premature isolation of the germ-plasm during ontogenesis nullified the influence on the development of the species of somatic modifications determined by the milieu. Albert Brachet, in his book *La vie créatrice des formes*, could write that "the milieu is not, properly speaking, an agent of formation, but rather of realization,"[34] invoking as an example the variety of forms of oceanic living beings within an identical milieu. And Maurice Caullery concluded his discussion in *Problème de l'évolution* by recognizing that evolution depends much more on the intrinsic properties of organisms than on the ambient milieu.[35]

But we know that the conception of the total autonomy of hereditary genetic material has been criticized. One critique emphasized that nucleoplasmatic disharmony tends to limit the hereditary omnipotence of genes. In sexual reproduction, although each parent supplies half of the genes, the mother supplies the egg cytoplasm. Now, the fact that offspring from the cross-breeding of two different species are not the same—depending on which of the species is the father or the mother—leads one to think that the genes' strength varies as a function of the cytoplasmic milieu. At the same time, H. Müller's experiments (1927) inducing mutations in fruit flies through the action of a milieu of penetrative radiation (X-rays), seemed to shed light on how an organic phenomenon that has perhaps been too smugly used to highlight the separation of the organism from the environment can be conditioned from the outside. Finally, there was a renewal of Lamarckism in the polemics—at least as ideological as scientific—surrounding the indignant repudiation of the "pseudo-science" of Russian geneticists, whom Trofim Lysenko led back to the "sound method" of Ivan Vladimirovich Michurin (1855–1935). Experiments on the vernalization of cultivated plants such as wheat and rye led Lysenko to affirm that hereditary modifications can be obtained and reinforced by variations in conditions of nutrition, maintenance, and climate, leading to a dislocation or rupture of the hereditary constitution of the organism, wrongly supposed by geneticists to be stable. Insofar as we can summarize the complex experimental facts within our present scope, we can say that, according to Lysenko, heredity is dependent on metabolism and metabolism is dependent on conditions of existence. Heredity would thus be the assimilation, by the living, over the course of succeeding generations, of exterior conditions. The ideological commentaries surrounding these facts and this theory do indeed bring to light its sense, regardless of its ability to accommodate, or even to withstand, the experimental counter-proofs and criticisms that are the rule in scientific discussion and that, of course, lie outside our competence.[36] It seems that the technical—that is, agronomic—aspect of the problem is essential. The Mendelian theory of heredity, by establishing the spontaneous character of mutations, tends to damp human—and specifically Soviet—ambitions for the total domination of nature and to limit the possibility of intentionally altering living species. Finally, and above all, recognition of the milieu's determining action has a political and social impact: it authorizes man's unlimited action on himself via the intermediary of the milieu. It offers hope for an experimental renewal of human nature. It thus appears progressive in the highest degree. Theory and practice are inseparable, as befits Marxist-Leninist dialectics. One can then understand how it is that genetics could be charged with all the sins of racism and slavery, and Mendel presented as the head of a retrograde, capitalist, and idealist biology.

It is clear that, although the heredity of acquired characteristics may have regained favor, this does not authorize one to designate the recent theories of Soviet biologists as Lamarckian without qualification. What is essential in Lamarck's ideas, as we have seen, is that the organism's adaptation to its milieu is attributed to the initiative of the organism's needs, efforts, and continual reactions. The milieu provokes the organism to orient its becoming by itself. Biological response by far exceeds physical stimulation. By rooting the phenomena of adaptation in need, which is at once pain and impatience, Lamarck centered the indivisible totality of the organism and the milieu on the point where life coincides with its own sense, where, through its sensibility, the living situates itself absolutely, either positively or negatively, within existence.

In Lamarck, as in the first theoreticians of the milieu, the notions of "circumstances" and "ambience" had a very different meaning from that in ordinary language. They evoked a spherical, centered arrangement. The terms *influences* and *influencing circumstances*, which Lamarck also used, take their meaning from astrological conceptions. When Buffon, in *De la dégéneration des animaux*, speaks of the "dye" from the sky, which man gradually receives, he uses, no doubt unconsciously, a term borrowed from Paracelsus.[37] The very notion of "climate" is, in the eighteenth century[38] as well as at the beginning of the nineteenth, an undivided notion, at once geographical, astronomical, and astrological. The climate is the change in the sky's appearance, degree by degree, from the equator to each pole, and it is also the influence that the sky exerts on the Earth.

We have already indicated that, in the beginning, the biological notion of the milieu combined an anthropo-geographical component with a mechanical one. The anthropo-geographical component was even, in a sense, the entirety of the notion, for it included the astronomical component, which Newton had converted into a notion of celestial mechanics. At its origin, geography was, for the Greeks, the projection of the heavens onto the earth, the bringing into correspondence of the sky and the earth: a correspondence at once topographical (geometry with cosmography) and hierarchical (physics and astrology). The co-ordination of the parts of the earth, and the subordination to the sky of an earth whose area is coordinated, were underlain by an astro-biological intuition of the Cosmos. Greek geography had its philosophy—that of the Stoics.[39] The intellectual relations between Posidonius on the one hand and Hipparches, Strabo, and Ptolemy on the other, are incontestable. What gives meaning to the geographical theory of milieu is the theory of universal sympathy, a vitalist intuition of universal determinism. This theory implies the comparison of the totality of things to an organism and the representation of this totality in the form of a sphere, centered on the situation of a privileged living being: man. This biocentric conception of the Cosmos persisted through the Middle Ages and blossomed in the Renaissance.

We know what became of the idea of the Cosmos with Copernicus, Kepler, and Galileo, and how dramatic the conflict was between the organic conception of the world and the conception of a universe decentered in relation to the ancient world's privileged center of reference, the land of living beings and man. From Galileo and Descartes on, one had to choose between two theories of milieu, that is, between two theories of space: a centered, qualified space, where the mi-*lieu* is a center; or a decentered, homogenous space, where the *mi*-lieu is an intermediary field. Pascal's famous text *Disproportion of Man* clearly shows the ambiguity of this term for a mind which cannot or does not want to choose between the need for existential security and the demands of scientific knowledge.[40] Pascal knows perfectly well that the Cosmos has broken to pieces, but the eternal silence of infinite spaces terrifies him. Man is no longer in the middle [*milieu*] of the world, but *he is a milieu* (a milieu between two infinities, a milieu between nothing and everything, a milieu between two extremes);[41] the milieu is *the state in which nature has placed us; we are floating on a vast milieu; man is in proportion with parts of the world, he has a relation to all that he knows*: "He needs space to contain him, time to exist in, motion to be alive, elements to constitute him, warmth and food for nourishment, air to breathe. He sees light, he feels bodies, everything in short is related to him."[42] We thus see three meanings of *milieu* intervene here: medial situation, fluid of sustenance, and vital environment. In developing the last sense of the term, Pascal presents his organic conception of the world, a return to Stoicism beyond and against Descartes:

> Since all things are both caused and causing, assisted and assisting, mediate and immediate, providing mutual support in a chain linking together naturally and imperceptibly the most distant and different things, I consider it as impossible to know the parts without knowing the whole as to know the whole without knowing the individual parts.[43]

And when Pascal defines the universe as an "infinite sphere whose center is everywhere and circumference nowhere,"[44] he paradoxically attempts, by using an image borrowed from the theosophical tradition, to reconcile the new scientific conception (which makes the universe an indefinite and undifferentiated milieu) with the ancient cosmological vision (which makes the world a finite totality connected to its center). The image Pascal uses here is a permanent myth of mystical thought, a myth of Neo-Platonic origin, in which the intuition of a spherical world centered on and by the living is combined with the already heliocentric cosmology of the Pythagoreans.[45]

Up to and including Newton, there was no one who did not take from Jacob Boehme, Henry More ("the Platonist of Cambridge"), and their Neo-Platonist cosmology some symbolic representation of what a ubiquitous action radiating out from a center would be. Newtonian space and ether maintain an absolute quality, which the scholars of the eighteenth and nineteenth centuries were not able to recognize: space, as the means for God's omnipresence, and ether, as the support and vehicle of forces. Newtonian science, which was to underlie so many empiricist and relativist professions of faith, is founded on metaphysics. Its empiricism masks its theological foundations. And in this way, the natural philosophy at the origin of the positivist and mechanicist conception of the milieu is in fact itself supported by the mystical intuition of a sphere of energy whose central action is identically present and effective at all points.[46]

If today it seems completely normal to anyone trained in mathematics or physics that the ideal of the objectivity of knowledge demands a decentering of the vision of things, it also seems that the moment has come to understand that in biology, following the words of J. S. Haldane in *The Philosophy of a Biologist*, "it is physics that is not an exact science." As Edouard Claparède writes: "What distinguishes the animal is the fact that it is a center in relation to those ambient forces which are, in relation to it, no more than stimulants or signals; a center, that is to say, a system of internal regulation, whose reactions are determined by an internal cause: the momentary need."[47] In this sense, the milieu on which the organism depends is structured, organized, by the organism itself. What the milieu offers the living is a function of demand. It is for this reason that, within what appears to man as a single milieu, various living beings carve out their specific and singular milieu in incomparable ways. Moreover, as a living being, man does not escape from the general law of living beings. The milieu proper to man is the world of his perception—in other words, the field of his pragmatic experience, the field in which his actions, oriented and regulated by the values immanent to his tendencies, pick out quality-bearing objects and situate them in relation to each other and to him. Thus the environment to which he is supposed to react is originally centered on him and by him.

Yet man as scientist and bearer of knowledge constructs a universe of phenomena and laws that he holds to be an absolute universe. The essential function of science is to devalorize the qualities of objects that comprise the milieu proper to man; science presents itself as the general theory of a real, that is to say, inhuman milieu. Sensory data are disqualified, quantified, identified. The imperceptible is presumed, and then detected and proven. Measurements substitute for appreciations, laws for habits, causality for hierarchy, and the objective for the subjective.

Hence the universe of the scientist [*l'homme savant*]. Einstein's physics is its ideal representation: a universe whose fundamental equations of intelligibility are the same, no matter what the system of reference may be. Because this universe maintains a direct relation to the milieu proper to living man—albeit a relation of negation and reduction—it confers upon this proper milieu a sort of privilege over the milieus proper to other living beings. Despite finding his ordinary perceptual experience contradicted and corrected by scientific research, living man [*l'homme vivant*] draws from his relation to the scientist [*l'homme savant*] a sort of unconscious self-conceit, which makes him prefer his own milieu over the milieus of other living beings, as having more reality and not just a different value. In fact, as a proper milieu for comportment and life, the milieu of man's sensory and technical values does not in itself have more reality than the milieus proper to the woodlouse or the gray mouse. In all rigor, the qualification *real* can only be applied to the absolute universe, the universal milieu of elements and movements disclosed by science. Its recognition as real is necessarily accompanied by the disqualification, as illusions or vital errors, of all subjectively centered proper milieus, including that of man.

The claim of science to dissolve living beings, which are centers of organization, adaptation, and invention, into the anonymity of the mechanical, physical, and chemical environment must be integral—that is, it must encapsulate the human living himself. We know well that this project did not appear too audacious to many scientists. But we must then ask, from a philosophical point of view, whether the origin of science does not reveal its meaning better than the claims of certain scientists do. In a humanity to which, from the scientific and even the materialist point of view, innate knowledge is rightly refused, the birth, becoming, and progress of science must be understood as a sort of enterprise as adventurous as life. Otherwise, one would have to admit the absurdity that reality contains the science of reality beforehand, as a part of itself. And we would then have to wonder to which among the needs of reality this ambition to determine reality scientifically could correspond.

But if science is the work of a humanity rooted in life before being enlightened by knowledge, if science is a fact in the world at the same time as it is a vision of the world, then it maintains a permanent and obligatory relation with perception. And thus the milieu proper to men is not situated within the universal milieu as contents in a container. A center does not resolve into its environment. A living being is not reducible to a crossroads of influences. From this stems the insufficiency of any biology that, in complete submission to the spirit of the physico-chemical sciences, would seek to eliminate all consideration of sense from its domain. From the biological and psychological point of view, a sense is an appreciation of values in relation to a need. And for the one who experiences and lives it, a need is an irreducible, and thereby absolute, system of reference.

Ethics: Truth and Subjectivity

The Hermeneutics of the Subject

Michel Foucault

Michel Foucault (1926–1984) was a philosopher and historian and one of the most original and dynamic thinkers of the postwar period. The lecture presented here was delivered as part of a course he offered in 1981–1982 at the illustrious Collège de France. In the lecture Foucault sketches an outline of the relation between what he calls "practices of the care of the self" and the capacity to seek and to know truth, from the early Greek and antique Roman philosophers to early Christian theologians. His evocative proposal that the task of thinking in late antiquity was to transform the thinker through ethical practice has had a major impact on contemporary anthropology.

This year I thought of trying the following arrangements:[1] I will lecture for two hours, from 9:15 until 11:15, with a short break of a few minutes after an hour to allow you to rest, or to leave if you are bored, and also to give me a bit of a rest. As far as possible I will try nevertheless to vary the two hours. That is to say, in the first hour, or at any rate in one of the two hours, I will give a somewhat more, let's say, theoretical and general exposition, and then, in the other hour, I will present something more like a textual analysis with, of course, all the obstacles and drawbacks of this kind of approach due to the fact that we cannot supply you with the texts and do not know how many of you there will be, etcetera. Still, we can always try. If it does not work we will try to find another method next year, or even this year. Does it bother you much to come at 9:15? No? It's okay? You are more fortunate than me, then.

Last year I tried to get a historical reflection underway on the theme of the relations between subjectivity and truth.[2] To study this problem I took as a privileged example, as a refracting surface if you like, the question of the regimen of sexual behavior and pleasures in Antiquity, the regimen of the *aphrodisia* you recall, as it appeared and was defined in the first two centuries AD.[3] It seemed to me that one of the interesting dimensions of this regimen was that the basic framework of modern European sexual morality was to be found in this regimen of the *aphrodisia*, rather than in so-called Christian morality, or worse, in so-called Judeo-Christian morality.[4] This year I would like to step back a bit from this precise example, and from the sexual material concerning the *aphrodisia* and sexual behavior, and extract from it the more general terms of the problem of "the subject and truth." More precisely, while I do not want in any way to eliminate or nullify the historical dimension in which I tried to situate this problem of subjectivity/truth relations, I would, however, like to present it in a much more general form. The question I would like to take up this year is this: In what historical form do the relations between the "subject" and "truth," elements that do not usually fall within the historian's practice or analysis, take shape in the West?

So, to start with I would like to take up a notion about which I think I said a few words last year.[5] This is the notion of "care of oneself." This is the best translation I can offer for a very complex, rich, and frequently employed Greek notion which had a long life throughout Greek culture: the notion of *epimeleia heautou*, translated into Latin with, of course, all the flattening of meaning which has so often been denounced or, at any rate, pointed out,[6] as *cura sui*.[7] *Epimeleia heautou* is care of oneself, attending to oneself, being concerned about oneself, etcetera. You will no doubt say that in order to study the relations between the subject and truth it is a bit paradoxical and rather artificial to select this notion of *epimeleia heautou*, to which the historiography of philosophy has not attached much importance hitherto. It is somewhat paradoxical and artificial to select this notion when everyone knows, says, and repeats, and has done so for a long time, that the question of the subject (the question of knowledge of the subject, of the subject's knowledge of himself) was originally posed in a very different expression and a very different precept: the famous Delphic prescription of *gnōthi seauton* ("know yourself").[8] So, when everything in the history of philosophy—and more broadly in the history of Western thought—tells us that the *gnōthi seau-*

ton is undoubtedly the founding expression of the question of the relations between the subject and truth, why choose this apparently rather marginal notion—that of the care of oneself, of *epimeleia heautou*—which is certainly current in Greek thought, but which seems not to have been given any special status? So, in this first hour I would like to spend some time on this question of the relations between the *epimeleia heautou* (care of the self) and the *gnōthi seauton* ("know yourself").

Relying on the work of historians and archeologists, I would like to make this very simple preliminary remark with regard to the "know yourself." We should keep the following in mind: In the glorious and spectacular form in which it was formulated and engraved on the temple stone, the *gnōthi seauton* originally did not have the value it later acquired. You know (and we will have to come back to this) the famous text in which Epictetus says that the precept "*gnōthi seauton*" was inscribed at the center of the human community.[9] In fact it undoubtedly was inscribed in this place, which was a center of Greek life, and later of the human community,[10] but it certainly did not mean "know yourself" in the philosophical sense of the phrase. The phrase did not prescribe self-knowledge, neither as the basis of morality, nor as part of a relationship with the gods. A number of interpretations have been suggested. There is Roscher's old interpretation, put forward in 1901 in an article in *Philologus*,[11] in which he recalled that the Delphic precepts were after all addressed to those who came to consult the god and should be read as kinds of ritual rules and recommendations connected with the act of consultation itself. You know the three precepts. According to Roscher, the precept *mēden agan* ("not too much") certainly does not designate or express a general ethical principle and measure for human conduct. *Mēden agan* ("not too much") means: You who have come to consult, do not ask too many questions, ask only useful questions and those that are necessary. The second precept concerning the *egguē* (the pledges)[12] would mean precisely this: When you consult the gods, do not make vows and commitments that you will not be able to honor. As for the *gnōthi seauton*, according to Roscher it would mean: When you question the oracle, examine yourself closely and the questions you are going to ask, those you wish to ask, and, since you must restrict yourself to the fewest questions and not ask too many, carefully consider yourself and what you need know. Defradas gives a much more recent interpretation, in 1954, in his book on *Les thèmes de la propagande delphique*.[13]

Defradas proposes a different interpretation, but which also shows, or suggests, that the *gnōthi seauton* is definitely not a principle of self-knowledge. According to Defradas, the three Delphic precepts were general demands for prudence: "not too much" in your requests and hopes and no excess in how you conduct yourself. The "pledges" was a precept warning those consulting against excessive generosity. As for the "know yourself," this was the principle [that] you should always remember that you are only a mortal after all, not a god, and that you should neither presume too much on your strength nor oppose the powers of the deity.

Let us skip this quickly. I want to stress something else which has much more to do with the subject with which I am concerned. Whatever meaning was actually given and attached to the Delphic precept "know yourself" in the cult of Apollo, it seems to me to be a fact that when this Delphic precept, this *gnōthi seauton*, appears in philosophy, in philosophical thought, it is, as we know, around the character of Socrates. Xenophon attests to this in the *Memorabilia*,[14] as does Plato in a number of texts to which we will have to return. Now not always, but often, and in a highly significant way, when this Delphic precept (this *gnōthi seauton*) appears, it is coupled or twinned with the principle of "take care of yourself" (*epimeleia heautou*). I say "coupled," "twinned." In actual fact, it is not entirely a matter of coupling. In some texts, to which we will have to return, there is, rather, a kind of subordination of the expression of the rule "know yourself" to the precept of care of the self. The *gnōthi seauton* ("know yourself") appears, quite clearly and again in a number of significant texts, within the more general framework of the *epimeleia heautou* (care of oneself) as one of the forms, one of the consequences, as a sort of concrete, precise, and particular application of the general rule: You must attend to yourself, you must not forget yourself, you must take care of yourself. The rule "know yourself" appears and is formulated within and at the forefront of this care. Anyway, we should not forget that in Plato's too well-known but still fundamental text, the *Apology*, Socrates appears as the person whose essential, fundamental, and original function, job, and position is to encourage others to attend to themselves, take care of themselves, and not neglect themselves. There are in fact three texts, three passages in the *Apology* that are completely clear and explicit about this.

The first passage is found in 29d of the *Apology*.[15] In this passage, Socrates, defending himself, making a kind of imaginary defense plea before his accusers and judges, answers the following objection. He is reproached with having ended up in a situation of which "he should be ashamed." The accusation, if you like, consists in saying: I am not really sure what evil you have done, but I avow all the same that it is shameful to have led the kind of life that results in you now finding yourself accused before the courts and in danger of being condemned, perhaps condemned to death. Isn't this, in the end, what is shameful, that someone has led a certain life, which while we do not know what it is, is such that he is in danger of being condemned to death by such a judgment? In this passage, Socrates replies that, on the contrary, he is very proud of having led this life and that if ever he was asked to lead a different life he would refuse. So: I am so proud of the life I have led that I would not change it even if you offered to acquit me. Here are Socrates' words: "Athenians, I am grateful to you and love you, but I shall obey God rather than you, and be sure that I will not stop practicing philosophy so long as I have breath and am able to, [exhorting] you and telling whoever I meet what they should do."[16] And what advice would he give if he is not condemned, since he had already given it before he was accused? To those he meets he will say, as he is accustomed to saying: "Dear friend, you are an Athenian, citizen of the greatest city, more famous than any other for its knowledge and might, yet are you not ashamed for devoting all your care (*epimeleisthai*) to increasing your wealth, reputation and honors while not caring for or even considering (*epimelē, phrontizeis*) your reason, truth and the constant improvement of your soul?" Thus Socrates recalls what he has always said and is quite determined to continue to say to those he will meet and stop to question: You care for a whole range of things, for your wealth and your reputation. You do not take care of yourself. He goes on: "And if anyone argues and claims that he does care [for his soul, for truth, for reason; M.F.], don't think that I shall let him go and go on my way: No, I shall question him, examine him and argue with him at length . . ."[17] Whoever I may meet, young or old, stranger or fellow citizen, this is how I shall act, and especially with you my fellow citizens, since you are my kin. For you should understand that this is what the god demands, and I believe that nothing better has befallen this city than my zeal in executing this command."[18] This "command," then, is the command by which the gods have entrusted Socrates with the task of stopping people, young and old, citizens or strangers, and saying to them: Attend to yourselves. This is Socrates' task.

In the second passage, Socrates returns to this theme of the care of the self and says that if the Athenians do in fact condemn him to death then he,

Socrates, will not lose a great deal. The Athenians, however, will suffer a very heavy and severe loss.[19] For, he says, there will no longer be anyone to encourage them to care for themselves and their own virtue unless the gods care enough about them to send someone to replace him, someone who will constantly remind them that they must be concerned about themselves.[20]

Finally, in 36b–c, there is the third passage, which concerns the penalty incurred. According to the traditional legal forms,[21] Socrates himself proposes the penalty he will accept if condemned. Here is the text: "What treatment do I deserve, what amends must I make for thinking I had to relinquish a peaceful life and neglect what most people have at heart—wealth, private interest, military office, success in the assembly, magistracies, alliances and political factions; for being convinced that with my scruples I would be lost if I followed such a course; for not wanting to do what was of no advantage either to you or myself; for preferring to do for each particular individual what I declare to be the greatest service, trying to persuade him to care (*epimelētheiē*) less about his property than about himself so as to make himself as excellent and reasonable as possible, to consider less the things of the city than the city itself, in short, to apply these same principles to everything? What have I deserved, I ask, for having conducted myself in this way [and for having encouraged you to attend to yourselves? Not punishment, to be sure, not chastisement, but; M.F.] something good, Athenians, if you want to be just."[22]

I will stop there for the moment. I just wanted to draw your attention to these passages, in which Socrates basically appears as the person who encourages others to care for themselves, and I would like you to note three or four important things. First, this activity of encouraging others to care for themselves is Socrates' activity, but it is an activity entrusted to him by the gods. In acting in this way Socrates does no more than carry out an order, perform a function or occupy a post (he uses the term *taxis*)[23] determined for him by the gods. In this passage you will also have been able to see that it is because the gods care for the Athenians that they sent Socrates, and may possibly send someone else, to encourage them to care for themselves.

Second, you see as well, and this is very clear in the last passage I read to you, that if Socrates cares for others, then this obviously means that he will not care for himself, or at any rate, that in caring for others he will neglect a range of other activities that are generally thought to be self-interested, profitable, and advantageous. So as to be able to care for others, Socrates has neglected his wealth and a number of civic advantages, he has renounced any political career, and he has not sought any office or magistracy. Thus the problem arises of the relation between the "caring for oneself" encouraged by the philosopher, and what caring for himself, or maybe sacrificing himself, must represent for the philosopher, that is to say, the problem, consequently, of the position occupied by the master in this matter of "caring for oneself."

Third, I have not quoted this passage at great length, but it doesn't matter, you can look it up: in this activity of encouraging others to attend to themselves Socrates says that with regard to his fellow citizens his role is that of someone who awakens them.[24] The care of the self will thus be looked upon as the moment of the first awakening. It is situated precisely at the moment the eyes open, when one wakes up and has access to the first light of day. This is the third interesting point in this question of "caring for oneself."

Finally, again at the end of a passage I did not read to you, there is the famous comparison of Socrates and the horsefly, the insect that chases and bites animals, making them restless and run about.[25] The care of oneself is a sort of thorn which must be stuck in men's flesh, driven into their existence, and which is a principle of restlessness and movement, of continuous concern throughout life. So I think this question of the *epimeleia heautou* should be rescued from the prestige of the *gnōthi seauton* that has somewhat overshadowed its importance. In a text, then, which I will try to explain to you a bit more precisely in a moment (the whole of the second part of the famous *Alcibiades*), you will see how the *epimeleia heautou* (the care of the self) is indeed the justificatory framework, ground, and foundation for the imperative "know yourself." So, this notion of *epimeleia heautou* is important in the figure of Socrates, with whom one usually associates, if not exclusively then at least in a privileged fashion, the *gnōthi seauton*. Socrates is, and always will be, the person associated with care of the self. In a series of late texts, in the Stoics, in the Cynics, and especially in Epictetus,[26] you will see that Socrates is always, essentially and fundamentally, the person who stops young men in the street and tells them: "You must care about yourselves."

The third point concerning this notion of *epimeleia heautou* and its connections with the *gnōthi seauton* is that the notion of *epimeleia heautou* did not just accompany, frame, and found the necessity of knowing oneself, and not solely when this necessity appeared in the thought, life, and figure of Socrates. It seems to me that the *epimeleia heautou* (the care of the self and the rule associated with it) remained a fundamental principle for describing the philosophical attitude throughout Greek, Hellenistic, and Roman culture. This notion of the care of the self was, of course, important in Plato. It was important for the Epicureans, since in Epicurus you find the frequently repeated expression: Every man should take care of his soul day and night and throughout his life.[27] For "take care of" Epicurus employs the verb *therapeuein*[28] which has several meanings: *therapeuein* refers to medical care (a kind of therapy for the soul which we know was important for the Epicureans),[29] but *therapeuein* is also the service provided by a servant to his master. You know also that *therapeuein* is related to the duties of worship, to the statutory regular worship rendered to a deity or divine power. The care of the self is crucially important in the Cynics. I refer, for example, to the text cited by Seneca in the first paragraphs of book seven of *De Benificiis*, in which the Cynic Demetrius, on the basis of a number of principles to which we will have to return because this is very important, explains how it is pointless to concern oneself with speculations about certain natural phenomena (like, for example, the origin of earthquakes, the causes of storms, the reason for twins), and that one should look instead to immediate things concerning oneself and to a number of rules by which one conducts oneself and controls what one does.[30] I don't need to tell you that the *epimeleia heautou* is important in the Stoics; it is central in Seneca with the notion of *cura sui* and it permeates the *Discourses* of Epictetus. Having to care about oneself is not just a condition for gaining access to the philosophical life, in the strict and full sense of the term. You will see, I will try to show you, how generally speaking the principle that one must take care of oneself became the principle of all rational conduct in all forms of active life that would truly conform to the principle of moral rationality. Throughout the long summer of Hellenistic and Roman thought, the exhortation to care for oneself became so widespread that it became, I think, a truly general cultural phenomenon.[31] What I would like to show you, what I would like to speak about this year, is this history that made this general cultural phenomenon (this exhortation, this general acceptance of the principle that one should take care of oneself) both a general cultural phenomenon peculiar to Hellenistic and Roman society (anyway, to its elite), and at the same time an event in thought.[32] It seems to me that the stake, the challenge for any history of thought, is precisely that of grasping when a cultural phenomenon of a determinate scale actually constitutes within the history of thought a decisive moment that is still significant for our modern mode of being subjects.

One word more: If this notion of the care of oneself, which we see emerging quite explicitly and clearly in the figure of Socrates, traversed and permeated ancient philosophy up to the threshold of Christianity, well, you will find this notion of *epimeleia* (of care) again in Christianity, or in what, to a certain extent, constituted its environment and preparation: Alexandrian spirituality. At any rate, you find this notion of given a particular meaning in Philo (*De Vita Contemplativa*).[33] You find it in Plotinus, in *Ennead*, II.[34] You find this notion of *epimeleia* also and especially in Christian asceticism: in Methodius of Olympus[35] and Basil of Caesarea.[36] It appears in Gregory of Nyssa: in *The Life of Moses*,[37] in the text on *The Song of Songs*,[38] and in the *Beatitudes*.[39] The notion of care of the self is found especially in Book XIII of *On Virginity*,[40] the title of which is, precisely, "That the care of oneself begins with freedom from marriage."[41] Given that, for Gregory of Nyssa, freedom from marriage (celibacy) is actually the first form, the initial inflection of the ascetic life, the assimilation of the first form of the care of oneself and freedom from marriage reveals the extent to which the care of the self had become a kind of matrix of Christian asceticism. You can see that the notion of *epimeleia heautou* (care of oneself) has a long history extending from the figure of Socrates stopping young people to tell them to take care of themselves up to Christian asceticism making the ascetic life begin with the care of oneself.

It is clear that in the course of this history the notion becomes broader and its meanings are both multiplied and modified. Since the purpose of this year's course will be to elucidate all this (what I am saying now being only a pure schema, a preliminary overview), let's say that within this notion of *epimeleia heautou* we should bear in mind that there is:

First, the theme of a general standpoint, of a certain way of considering things, of behaving in the world, undertaking actions, and having relations with other people. The *epimeleia heautou* is an attitude towards the self, others, and the world;

Second, the *epimeleia heautou* is also a certain form of attention, of looking. Being concerned about oneself implies that we look away from the outside to . . . I was going to say "inside." Let's leave to one side this word, which you can well imagine raises a host of problems, and just say that we must convert our looking from the outside, from others and the world etc., towards "oneself." The care of the self implies a certain way of attending to what we think and what takes place in our thought. The word *epimeleia* is related to *meletē*, which means both exercise and meditation.[42] Again, all this will have to be elucidated;

Third, the notion of *epimeleia* does not merely designate this general attitude or this form of attention turned on the self. The *epimeleia* also always designates a number of actions exercised on the self by the self, actions by which one takes responsibility for oneself and by which one changes, purifies, transforms, and transfigures oneself. It involves a series of practices, most of which are exercises that will have a very long destiny in the history of Western culture, philosophy, morality, and spirituality. These are, for example, techniques of meditation,[43] of memorization of the past, of examination of conscience,[44] of checking representations which appear in the mind,[45] and so on.

With this theme of the care of the self, we have then, if you like, an early philosophical formulation, appearing clearly in the fifth century BC, of a notion which permeates all Greek, Hellenistic, and Roman philosophy, as well as Christian spirituality, up to the fourth and fifth centuries AD. In short, with this notion of *epimeleia heautou* we have a body of work defining a way of being, a standpoint, forms of reflection, and practices which make it an extremely important phenomenon not just in the history of representations, notions, or theories, but in the history of subjectivity itself or, if you like, in the history of practices of subjectivity. Anyway, as a working hypothesis at least, this one-thousand-year development from the appearance of the first forms of the philosophical attitude in the Greeks to the first forms of Christian asceticism—from the fifth century BC to the fifth century AD— can be taken up starting from this notion of *epimeleia heautou*. Between the philosophical exercise and Christian asceticism there are a thousand years of transformation and evolution in which the care of the self is undoubtedly one of the main threads or, at any rate, to be more modest, let's say one of the possible main threads.

Even so, before ending these general remarks, I would like to pose the following question: Why did Western thought and philosophy neglect the notion of *epimeleia heautou* (care of the self) in its reconstruction of its own history? How did it come about that we accorded so much privilege, value, and intensity to the "know yourself" and omitted, or at least, left in the shadow, this notion of care of the self that, in actual fact, historically, when we look at the documents and texts, seems to have framed the principle of "know yourself" from the start and to have supported an extremely rich and dense set of notions, practices, ways of being, forms of existence, and so on? Why does the *gnōthi seauton* have this privileged status for us, to the detriment of the care of oneself? Okay, what I will sketch out here are of course hypotheses with many question marks and ellipses.

Just to begin with, entirely superficially and without resolving anything, but as something that we should maybe bear in mind, I think we can say that there is clearly something a bit disturbing for us in this principle of the care of the self. Indeed, going through the texts, the different forms of philosophy and the different forms of exercises and philosophical or spiritual practices, we see the principle of care of the self expressed in a variety of phrases like: "caring for oneself," "taking care of the self," "withdrawing into oneself," "retiring into the self," "finding one's pleasure in oneself," "seeking no other delight but in the self," "remaining in the company of oneself," "being the friend of oneself," "being in one's self as in a fortress," "looking after" or "devoting oneself to oneself," "respecting oneself," etc. Now you are well aware that there is a certain tradition (or rather, several traditions) that dissuades us (us, now, today) from giving any positive value to all these expressions, precepts, and rules, and above all from making them the basis of a morality: All these injunctions to exalt oneself, to devote oneself to oneself, to turn in on oneself, to offer service to oneself, sound to our ears rather like—what? Like a sort of challenge and defiance, a desire for radical ethical change, a sort of moral dandyism, the assertion-challenge of a fixed aesthetic and individual stage.[46] Or else they sound to us like a somewhat melancholy and sad expression of the withdrawal of the individual who is unable to hold on to and keep firmly before his eyes, in his grasp and for himself, a collective morality (that of the city-state, for example), and who, faced with the disintegration of this collective morality, has naught else to do but attend to himself.[47] So, the immediate, initial connotations and overtones of all these expressions direct us away from thinking about these precepts in positive terms. Now, in all of the ancient thought I am talking about, whether it be

Socrates or Gregory of Nyssa, "taking care of oneself" always has a positive and never a negative meaning. A further paradox is that this injunction to "take care of oneself" is the basis for the constitution of what have without doubt been the most austere, strict, and restrictive moralities known in the West, moralities which, I repeat, should not be attributed to Christianity (this was the object of last year's course), but rather to the morality of the first centuries BC and the first centuries AD. (Stoic, Cynic, and, to a certain extent, Epicurean morality). Thus, we have the paradox of a precept of care of the self which signifies for us either egoism or withdrawal, but which for centuries was rather a positive principle that was the matrix for extremely strict moralities. A further paradox which should be mentioned to explain the way in which this notion of care of the self was somehow overshadowed is that the strict morality and austere rules arising from the principle "take care of yourself" have been taken up again by us: These rules in fact appear, or reappear, either in a Christian morality or in a modern, non-Christian morality: However, they do so in a different context. These austere rules, which are found again identical in their codified structure, appear reacclimatized, transposed, and transferred within a context of a general ethic of non-egoism taking the form either of a Christian obligation of self-renunciation or of a "modern" obligation towards others—whether this be other people, the collectivity, the class, or the fatherland etc. So, Christianity and the modern world has based all these themes and codes of moral strictness on a morality of non-egoism whereas in actual fact they were born within an environment strongly marked by the obligation to take care of oneself. I think this set of paradoxes is one of the reasons why this theme of the care of the self was somewhat neglected and able to disappear from the concerns of historians.

However, I think there is a reason that is much more fundamental than these paradoxes of the history of morality. This pertains to the problem of truth and the history of truth. It seems to me that the more serious reason why this precept of the care of the self has been forgotten, the reason why the place occupied by this principle in ancient culture for nigh on one thousand years has been obliterated, is what I will call—with what I know is a bad, purely conventional phrase—the "Cartesian moment." It seems to me that the "Cartesian moment," again within a lot of inverted commas, functioned in two ways. It came into play in two ways: by philosophically requalifying the *gnōthi seauton* (know yourself), and by discrediting the *epimeleia heautou* (care of the self).

First, the Cartesian moment philosophically requalified the *gnōthi seauton* (know yourself). Actually, and here things are very simple, the Cartesian approach, which can be read quite explicitly in the *Meditations*,[48] placed self-evidence (*l'évidence*) at the origin, the point of departure of the philosophical approach—self-evidence as it appears, that is to say as it is given, as it is actually given to consciousness without any possible doubt [. . .] The Cartesian approach [therefore] refers to knowledge of the self, as a form of consciousness at least. What's more, by putting the self-evidence of the subject's own existence at the very source of access to being, this knowledge of oneself (no longer in the form of the test of self-evidence, but in the form of the impossibility of doubting my existence as subject) made the "know yourself" into a fundamental means of access to truth. Of course, there is a vast distance between the Socratic *gnōthi seauton* and the Cartesian approach. However, you can see why, from the seventeenth century, starting from this step, the principle of *gnōthi seauton* as founding moment of the philosophical method was acceptable for a number of philosophical approaches or practices. But if the Cartesian approach thus requalified the *gnōthi seauton*, for reasons that are fairly easy to isolate, at the same time—and I want to stress this—it played a major part in discrediting the principle of care of the self and in excluding it from the field of modern philosophical thought.

Let's stand back a little to consider this. We will call, if you like, "philosophy" the form of thought that asks, not of course what is true and what is false, but what determines that there is and can be truth and falsehood and whether or not we can separate the true and the false. We will call "philosophy" the form of thought that asks what it is that enables the subject to have access to the truth and which attempts to determine the conditions and limits of the subject's access to the truth. If we call this "philosophy," then I think we could call "spirituality" the search, practice, and experience through which the subject carries out the necessary transformations on himself in order to have access to the truth. We will call "spirituality" then the set of these researches, practices, and experiences, which may be purifications, ascetic exercises, renunciations, conversions of looking, modifications of existence, etc., which are, not for knowledge but for the subject, for the subject's very being, the price to be paid for access to the truth. Let's say that spirituality, as it appears in the West at least, has three characteristics.

Spirituality postulates that the truth is never given to the subject by right. Spirituality postulates that the subject as such does not have right of access to the truth and is not capable of having access to the truth. It postulates that the truth is not given to the subject by a simple act of knowledge (*connaissance*), which would be founded and justified simply by the fact that he is the subject and because he possesses this or that structure of subjectivity. It postulates that for the subject to have right of access to the truth he must be changed, transformed, shifted, and become, to some extent and up to a certain point, other than himself. The truth is only given to the subject at a price that brings the subject's being into play. For as he is, the subject is not capable of truth. I think that this is the simplest but most fundamental formula by which spirituality can be defined. It follows that from this point of view there can be no truth without a conversion or a transformation of the subject. This conversion, this transformation of the subject—and this will be the second major aspect of spirituality—may take place in different forms. Very roughly we can say (and this is again a very schematic survey) that this conversion may take place in the form of a movement that removes the subject from his current status and condition (either an ascending movement of the subject himself, or else a movement by which the truth comes to him and enlightens him). Again, quite conventionally, let us call this movement, in either of its directions, the movement of *erōs* (love). Another major form through which the subject can and must transform himself in order to have access to the truth is a kind of work. This is a work of the self on the self, an elaboration of the self by the self, a progressive transformation of the self by the self for which one takes responsibility in a long labor of ascesis (*askēsis*). *Erōs* and *askēsis* are, I think, the two major forms in Western spirituality for conceptualizing the modalities by which the subject must be transformed in order finally to become capable of truth. This is the second characteristic of spirituality.

Finally, spirituality postulates that once access to the truth has really been opened up, it produces effects that are, of course, the consequence of the spiritual approach taken in order to achieve this, but which at the same time are something quite different and much more: effects which I will call "rebound" ("*de retour*"), effects of the truth on the subject. For spirituality, the truth is not just what is given to the subject, as reward for the act of knowledge as it were, and to fulfill the act of knowledge. The truth enlightens the subject; the truth gives beatitude to the subject; the truth gives the subject tranquility of the soul. In short, in the truth and in access to the truth, there is something that fulfills the subject himself, which fulfills or transfigures his very being. In short, I think we can say that in and of itself an act of knowledge could never give access to the truth unless it was prepared, accompanied, doubled, and completed by a certain transformation of the subject; not of the individual, but of the subject himself in his being as subject.

There is no doubt an enormous objection to everything I have been saying, an objection to which it will be necessary to return, and which is, of course, the gnosis.[49] However, the gnosis, and the whole Gnostic movement, is precisely a movement that overloads the act of knowledge (*connaissance*), to [which] sovereignty is indeed granted in access to the truth. This act of knowledge is overloaded with all the conditions and structure of a spiritual act. The gnosis is, in short, that which tends to transfer, to transpose, the forms and effects of spiritual experience into the act of knowledge itself. Schematically, let's say that throughout the period we call Antiquity, and in quite different modalities, the philosophical question of "how to have access to the truth" and the practice of spirituality (of the necessary transformations in the very being of the subject which will allow access to the truth), these two questions, these two themes, were never separate. It is clear they were not separate for the Pythagoreans. Neither were they separate for Socrates and Plato: the *epimeleia heautou* (care of the self) designates precisely the set of conditions of spirituality, the set of transformations of the self, that are the necessary conditions for having access to the truth. So, throughout Antiquity (in the Pythagoreans, Plato, the Stoics, Cynics, Epicureans, and Neo-Platonists), the philosophical theme (how to have access to the truth?) and the question of spirituality (what transformations in the being of the subject are necessary for access to the truth?) were never separate. There is, of course, the exception, the major and fundamental exception: that of the one who is called "the" philosopher,[50] because he was no doubt the only philosopher in Antiquity for whom the question of spirituality was least important; the philosopher whom we have recognized as the founder of philosophy in the modern sense of the term: Aristotle. But as everyone knows, Aristotle is not the pinnacle of Antiquity but its exception.

Now, leaping over several centuries, we can say that we enter the modern age (I mean, the history of truth enters its modern period) when it is assumed that what gives access to the truth, the condition for the subject's access to the truth, is knowledge (*connaissance*) and knowledge alone. It seems to me that what I have called the "Cartesian moment" takes on its position and meaning at this point, without in any way my wanting to say that it is a question of Descartes, that he was its inventor or that he was the first to do this. I think the modern age of the history of truth begins when knowledge itself and knowledge alone gives access to the truth. That is to say, it is when the philosopher (or the scientist, or simply someone who seeks the truth) can recognize the truth and have access to it in himself and solely through his activity of knowing, without anything else being demanded of him and without him having to change or alter his being as subject. Of course, this does not mean that the truth is obtained without conditions. But these conditions are of two orders, neither of which fall under the conditions of spirituality. On the one hand, there are the internal conditions of the act of knowledge and of the rules it must obey to have access to the truth: formal conditions, objective conditions, formal rules of method, the structure of the object to be known.[51] However, in any case, the conditions of the subject's access to the truth are defined within knowledge. The other conditions are extrinsic. These are conditions such as: "In order to know the truth one must not be mad" (this is an important moment in Descartes).[52] They are also cultural conditions: to have access to the truth we must have studied, have an education, and operate within a certain scientific consensus. And there are moral conditions: to know the truth we must make an effort, we must not seek to deceive our world, and the interests of financial reward, career, and status must be combined in a way that is fully compatible with the norms of disinterested research, etcetera. As you can see, these are all conditions that are either intrinsic to knowledge or extrinsic to the act of knowledge, but which do not concern the subject in his being; they only concern the individual in his concrete existence, and not the structure of the subject as such. At this point (that is, when we can say: "As such the subject is, anyway, capable of truth"—with the two reservations of conditions intrinsic to knowledge and conditions extrinsic to the individual), when the subject's being is not put in question by the necessity of having access to the truth, I think we have entered a different age of the history of relations between subjectivity and truth. And the consequence—or, if you like, the other aspect of this—is that access to truth, whose sole condition is henceforth knowledge, will find reward and fulfillment in nothing else but the indefinite development of knowledge. The point of enlightenment and fulfillment, the moment of the subject's transfiguration by the "rebound effect" on himself of the truth he knows, and which passes through, permeates, and transfigures his being, can no longer exist. We can no longer think that access to the truth will complete in the subject, like a crowning or a reward, the work or the sacrifice, the price paid to arrive at it. Knowledge will simply open out onto the indefinite dimension of progress, the end of which is unknown and the advantage of which will only ever be realized in the course of history by the institutional accumulation of bodies of knowledge, or the psychological or social benefits to be had from having discovered the truth after having taken such pains to do so. As such, henceforth the truth cannot save the subject. If we define spirituality as being the form of practices which postulate that, such as he is, the subject is not capable of the truth, but that, such as it is, the truth can transform and save the subject, then we can say that the modern age of the relations between the subject and truth begin when it is postulated that, such as he is, the subject is capable of truth, but that, such as it is, the truth cannot save the subject. Okay, a short rest if you like. Five minutes and then we will begin again.

I would like to say two or three more words because, despite my good intentions and a well-structured use of time, I have not entirely kept within the hour as I hoped. So I will say a few more words on this general theme of the relations between philosophy and spirituality and the reasons for the gradual elimination of the notion of care of the self from philosophical thought and concern. I was saying that it seemed to me that at a certain moment (and when I say "moment," there is absolutely no question of giving it a date and localizing or individualizing it around just one person) the link was broken, definitively I think, between access to the truth, which becomes the autonomous development of knowledge (*connaissance*), and the requirement of the subject's transformation of himself and of his being. When I say "I think it was definitively broken," I don't need to tell you that I don't believe any such thing, and that what is interesting is precisely that the links were not broken abruptly as if by the slice of a knife.

Let's consider things upstream first of all. The break does not occur just like that. It does not take place on the day Descartes laid down the rule of self-evidence or discovered the Cogito, etc. The work of disconnecting, on the one hand, the principle of an access to truth accomplished in terms of the knowing subject alone from, on the other, the spiritual necessity of the subject's work on himself, of his self-transformation and expectation of enlightenment and transfiguration from the truth, was underway long before. The dissociation had begun to take place long before and a certain wedge had been inserted between these two components. And of course, we should look for this wedge . . . in science? Not at all. We should look for it in theology (the theology which, precisely, with Aquinas, the scholastics, etc., was able to be founded on Aristotle—remember what I was just saying—and which will occupy the place we know it to have in Western reflection). This theology, by claiming, on the basis of Christianity of course, to be rational reflection founding a faith with a universal vocation, founded at the same time the principle of a knowing subject in general, of a knowing subject who finds both his point of absolute fulfillment and highest degree of perfection in God, who is also his Creator and so his model. The correspondence between an omniscient God and subjects capable of knowledge, conditional on faith of course, is undoubtedly one of the main elements that led Western thought—or its principal forms of reflection—and philosophical thought in particular, to extricate itself, to free itself, and separate itself from the conditions of spirituality that had previously accompanied it and for which the *epimeleia heautou* was the most general expression. I think we should be clear in our minds about the major conflict running through Christianity from the end of the fifth century—St. Augustine obviously—up to the seventeenth century. During these twelve centuries the conflict was not between spirituality and science, but between spirituality and theology. The best proof that it was not between spirituality and science is the blossoming of practices of spiritual knowledge, the development of esoteric knowledge, the whole idea—and it would be interesting to reinterpret the theme of Faust along these lines[53]—that there cannot be knowledge without a profound modification in the subject's being. That alchemy, for example, and a whole stratum of knowledge, was at this time thought to be obtainable only at the cost of a modification in the subject's being clearly proves that there was no constitutive or structural opposition between science and spirituality. The opposition was between theological thought and the requirement of spirituality. Thus the disengagement did not take place abruptly with the appearance of modern science. The disengagement, the separation, was a slow process whose origin and development should be located, rather, in theology.

Neither should we think that the break was made, and made definitively, at the moment I have called, completely arbitrarily, the "Cartesian moment." Rather, it is very interesting to see how the question of the relation between the conditions of spirituality and the problem of the development of truth and the method for arriving at it was posed in the seventeenth century. Take, for example, the very interesting notion that is typical of the end of the sixteenth and the beginning of the seventeenth century: the notion of "reform of the understanding."

Take, precisely, the first nine paragraphs of Spinoza's *Treatise on the Correction of the Understanding*.[54] You can see quite clearly there—and for well-known reasons that we don't need to emphasize—how in formulating the problem of access to the truth Spinoza linked the problem to a series of requirements concerning the subject's very being: In what aspects and how must I transform my being as subject? What conditions must I impose on my being as subject so as to have access to the truth, and to what extent will this access to the truth give me what I seek, that is to say the highest good, the sovereign good? This is a properly spiritual question, and the theme of the reform of the understanding in the seventeenth century is, I think, entirely typical of the still very strict, close, and tight links between, let's say, a philosophy of knowledge and a spirituality of the subject's transformation of his own being. If we now consider things downstream, if we cross over to the other side, starting with Kant, then here again we see that the structures of spirituality have not disappeared either from philosophical reflection or even, perhaps, from knowledge (*savoir*). There would be . . . but then I do not really want to outline it now, I just want to point out a few things. Read again all of nineteenth century philosophy—well, almost all: Hegel anyway, Schelling, Schopenhauer, Nietzsche, the Husserl of the *Krisis*[55] and Heidegger as well[56]—and you see precisely here also that knowledge (*connaissance*), the activity of knowing, whether [it] is discredited, devalued, considered critically, or rather, as in Hegel, exalted, is nonetheless still linked to the requirements of spirituality. In all these philosophies, a certain structure of spirituality tries to link knowledge, the activity of knowing, and the conditions and effects of this activity, to a transformation in the subject's being. *The Phenomenology of Mind*, after all, has no other meaning.[57] The entire history of nineteenth century philosophy can, I think, be thought of as a kind of pressure to try to rethink the structures of spirituality within a philosophy that, since Cartesianism, or at any rate since seventeenth century philosophy, tried to get free from these self-same structures. Hence the hostility, and what's more the profound hostility, of all the "classical" type of philosophers—all those who invoke the tradition of Descartes, Leibniz, etcetera—towards the philosophy of the nineteenth century that poses, at least implicitly, the very old question of spirituality and which, without saying so, rediscovers the care of the self.

However, I would say that this pressure, this resurgence, this reappearance of the structures of spirituality is nonetheless quite noticeable even within the field of knowledge (savoir) strictly speaking. If it is true, as all scientists say, that we can recognize a false science by the fact that access to it requires the subject's conversion and that it promises enlightenment for the subject at the end of its development; if we can recognize a false science by its structure of spirituality (which is self-evident; every scientist knows this), we should not forget that in those forms of knowledge (*savoir*) that are not exactly sciences, and which we should not seek to assimilate to the structure of science, there is again the strong and clear presence of at least certain elements, certain requirements of spirituality. Obviously, I don't need to draw you a picture: you will have immediately identified forms of knowledge like Marxism or psychoanalysis. It goes without saying that it would be completely wrong to identify these with religion. This is meaningless and contributes nothing. However, if you take each of them, you know that in both Marxism and psychoanalysis, for completely different reasons but with relatively homologous effects, the problem of what is at stake in the subject's being (of what the subject's being must be for the subject to have access to the truth) and, in return, the question of what aspects of the subject may be transformed by virtue of his access to the truth, well, these two questions, which are once again absolutely typical of spirituality, are found again at the very heart of, or anyway, at the source and outcome of both of these know ledges. I am not at all saying that these are forms of spirituality. What I mean is that, taking a historical view over some, or at least one or two millennia, you find again in these forms of knowledge the questions, interrogations, and requirements which, it seems to me, are the very old and fundamental questions of the *epimeleia heautou*, and so of spirituality as a condition of access to the truth. What has happened, of course, is that neither of these two forms of knowledge has openly considered this point of view clearly and willingly. There has been an attempt to conceal the conditions of spirituality specific to these forms of knowledge within a number of social forms. The idea of the effect of a class position or of the party, of allegiance to a group or membership of a school, of initiation or of the analyst's training, etc., all refer back to these questions of the condition of the subject's preparation for access to the truth, but conceived of in social terms, in terms of organization. They have not been thought of in terms of the historical thrust of the existence of spirituality and its requirements. Moreover, at the same time the price paid for transposing or reducing these questions of "truth and the subject" to problems of membership (of a group, a school, a party, a class, etc.), has been, of course, that the question of the relations between truth and the subject has been forgotten. The interest and force of Lacan's analyses seems to me to be due precisely to this: It seems to me that Lacan has been the only one since Freud who has sought to refocus the question of psychoanalysis on precisely this question of the relations between the subject and truth.[58] That is to say, in terms which are of course absolutely foreign to the historical tradition of this spirituality, whether of Socrates or Gregory of Nyssa and everyone in between, in terms of psychoanalytic knowledge itself, Lacan tried to pose what historically is the specifically spiritual question: that of the price the subject must pay for saying the truth, and of the effect on the subject of the fact that he has said, that he can and has said the truth about himself. By restoring this question I think Lacan actually reintroduced into psychoanalysis the oldest tradition, the oldest questioning, and the oldest disquiet of the *epimeleia heautou*, which was the most general form of spirituality. Of course, a question arises, which I will not answer, of whether psychoanalysis itself can, in its own terms, that is to say in terms of the effects of knowledge (*connaissance*), pose the question of the relations of the subject to truth, which by definition—from the point of view of spirituality, and anyway of the *epimeleia heautou*—cannot be posed in terms of knowledge (*connaissance*).

That is what I wanted to say about this. Now let's go on to a more simple exercise. Let's return to the texts. So, there is obviously no question of me rewriting the entire history of the notion, practice, and rules of the care of the self I have been referring to. This year, and once again subject to my sloppy timekeeping and inability to keep to a timetable, I will try to isolate three moments which seem to me to be interesting: the Socratic-Platonic moment, the appearance of the *epimeleia heautou* in philosophical reflection; second, the period of the golden age of the culture of the self, of the cultivation of oneself, of the care of oneself, which we can place in the first two centuries AD; and then, roughly, the transition from pagan philosophical ascesis to Christian asceticism in the fourth and fifth centuries.[59]

The first moment: Socratic-Platonic. Basically, then, the text I would like to refer to is the analysis, the theory itself of the care of the self; the extended theory developed in the second part, the conclusion, of the dialogue called *Alcibiades*. Before reading some of this text, I would like to recall two things. First, if it is true that the care of the self emerges in philosophical reflection with Socrates, and in the *Alcibiades* in particular, even so we should not forget that from its origin and throughout Greek culture the principle of "taking care of oneself"—as a rule and positive requirement from which a great deal is expected—was not an instruction for philosophers, a philosopher's interpellation of young people passing in the street. It is not an intellectual attitude; it is not advice given by wise old men to overeager young people. No, the assertion, the principle "one ought to take care of oneself," was an old maxim of Greek culture. In particular it was a Lacedaemonian maxim. In a text which, since it is from Plutarch, is fairly late, but which refers to what is clearly an ancestral and centuries-old saying, Plutarch reports a comment supposedly made by Anaxandridas, a Lacedaemonian, a Spartan, who is asked one day: You Spartans really are a bit strange. You have a lot of land and your territory is huge, or anyway substantial. Why don't you cultivate it yourselves, why do you entrust it to helots? And Anaxandridas is supposed to have answered: Well, quite simply, so that we can take care of ourselves.[60] Of course, when the Spartan says here: we have to take care of ourselves and so we do not have to cultivate our lands, it is quite clear that this has nothing to do [with philosophy]. In these people, for whom philosophy, intellectualism, etcetera, had no great positive value, taking care of themselves was the affirmation of a form of existence linked to a privilege, and to a political privilege: If we have helots, if we do not cultivate our lands ourselves, if we delegate all these material cares to others, it is so that we can take care of ourselves. The social, economic, and political privilege of this close-knit group of Spartan aristocrats was displayed in the form of: We have to look after ourselves, and to be able to do that we have entrusted our work to others. You can see then that "taking care of oneself'" is not at all philosophical but doubtless a fairly common principle linked, however, and we will find this again and again in the history of the *epimeleia heautou*, to a privilege, which in this case is political, economic, and social.

So when Socrates takes up and formulates the question of the *epimeleia heautou*, he does so on the basis of a tradition. Moreover, Sparta is referred to in the first major theory of the care of the self in the *Alcibiades*. So, let's move on now to this text, *Alcibiades*. Today, or next week, I will come back to the problems, not of its authenticity, which are more or less settled, but of its dating, which are very complicated.[61] But it is no doubt better to study the text itself and see the questions as they arise. I pass very quickly over the beginning of the dialogue of *Alcibiades*. I note only that right at the start we see Socrates accosting Alcibiades and remarking to him that until now he, Socrates, in contrast to Alcibiades' other lovers, has never approached Alcibiades, and that he has only decided to do so today. He has made up his mind to do so because he is aware that Alcibiades has something in mind.[62] He has something in mind, and Alcibiades is asked the old, classic question of Greek education, which goes back to Homer, etcetera:[63] Suppose you were offered the following choice, either to die today or to continue leading a life in which you will have no glory; which would you prefer? Well, [Alcibiades replies]: I would rather die today than lead a life that will bring me no more than what I have already. This is why Socrates approaches Alcibiades. What is it that Alcibiades has already and in comparison with which he wants something else? The particulars of Alcibiades' family, his status in the city, and his ancestral privileges place him above others. He has, the text says, "one of the most enterprising families of the city."[64] On his father's side—his father was a Eupatrid—he has connections, friends, and wealthy and powerful relatives. The same is true on the side of his mother, who was an Alcmaeonid.[65] Moreover, although he had lost both of his parents, his tutor was no nonentity, but Pericles. Pericles rules the roost in the city, even in Greece, and even in some barbarian countries.[66] Added to which, Alcibiades has a huge fortune. On the other hand, as everyone knows, Alcibiades is beautiful. He is pursued by numerous lovers and has so many and is so proud of his beauty and so arrogant that he has rejected all of them, Socrates being the only one who continues to pursue him. Why is he the only one?

He is the only one precisely because Alcibiades, by dint of having rejected all his lovers, has come of age. This is the famous critical age of boys I spoke about last year,[67] after which one can no longer really love them. However, Socrates continues to take an interest in Alcibiades. He continues to be interested in Alcibiades and even decides to speak to him for the first time. Why? Because, as I said to you a moment ago, he has clearly understood that Alcibiades has in mind something more than just benefiting from his connections, family, and wealth for the rest of his life, and as for his beauty, this is fading. Alcibiades does not want to be satisfied with this. He wants to turn to the people and take the city's destiny in hand: he wants to govern the others. In short, [he] is someone who wants to transform his statutory privilege and preeminence into political action, into his effective government of others. It is inasmuch as this intention is taking shape, at the point when Alcibiades—having taken advantage or refused to take advantage of others with his beauty—is turning to the government of others (after *erōs*, the *polis*, the city-state), that Socrates hears the voice of the god who inspires him to speak to Alcibiades. He has something to do: to transform statutory privilege and preeminence into the government of others. It is clear in the Alcibiades that the question of the care of the self arises at this point. The same thing can be found in what Xenophon says about Socrates. For example, in book III of the *Memorabilia*, Xenophon cites a dialogue, a meeting between Socrates and the young Charmides.[68] Charmides is also a young man on the threshold of politics, no doubt a little older than the Alcibiades of Plato's text since he is already mature enough to participate in the Assembly and give his views. Except that the Charmides who is heard in the Assembly, who gives his views and whose views are listened to because they are wise, is shy. He is shy, and although he is listened to and knows that everyone listens to him when considering things in a small group, he shrinks from speaking in public. And it is about this that Socrates says to him: Even so, you should pay heed to yourself; apply your mind to yourself, be aware of your qualities and in this way you will be able to participate in political life. He does not use the expression *epimeleia heautou* or *epimelei sautou*, but the expression "apply your mind." *Noūn prosekhei:*[69] apply your mind to yourself. But the situation is the same. It is the same, but reversed: Charmides, who despite his wisdom dares not enter political activity, must be encouraged, whereas with Alcibiades we are dealing with a young man champing at the bit, who only asks to enter politics and to transform his statutory advantages into real political action.

Now, asks Socrates, and this is where the part of the dialogue I want to study more closely begins, if you govern the city, if you are to be able to govern it, you must confront two sorts of rivals.[70] On the one hand there are the internal rivals you will come up against in the city, because you are not the only one who wants to govern. And then, when you are governing them, you will come up against the city's enemies. You will come up against Sparta and the Persian Empire. Now, says Socrates, you know very well how it is with both the Lacedaemonians and the Persians: they outmatch Athens and you. In wealth first of all: However wealthy you may be, can you compare your wealth to that of the Persian King? As for education, can you really compare your education with that of the Lacedaemonians and Persians? There is a brief description of Spartan education, which is not put forward as a model but as a mark of quality at least; an education that ensures firmness, greatness of soul, courage, endurance, the taste for victory and honor, etcetera. Persian education, and the passage here is interesting, also has great advantages. In the education given to the King, from the earliest age—in short, from when he is old enough to understand—the young prince is surrounded by four teachers: one is the teacher of wisdom (*sophia*), another of justice (*dikaiosunē*), the third a master of temperance (*sōphrosunē*), and the fourth a master of courage (*andreia*). With regard to the date of the text, the first problem to reckon with is the following: on the one hand, as you know, fascination and interest in Sparta is constant in Plato's dialogues, starting with the Socratic dialogues; however, the interest in and fascination with Persia is something which is thought to appear late in Plato and the Platonists [. . .]. How then has Alcibiades been trained in comparison with this education, whether Spartan or Persian? Well, says Socrates, consider what has happened. After the death of your parents you were entrusted to Pericles. For sure, Pericles "may lord it over his city, Greece and some barbarian States." However, in the event, he could not educate his sons. He had two of them, both good for nothing. Consequently you have come out badly. But one should not count on a serious training from this direction. And then again, your tutor Pericles entrusted you to an old slave (Zopyrus the Thracian) who was a monument to ignorance and so had nothing to teach you. Under these conditions, Socrates says to Alcibiades, you should make a little comparison: you want to enter political life, to take the destiny of the city in hand, and you do not have the wealth of your rivals, and above all you do not have their education. You should take a bit of a look at yourself, you should know yourself. And we see appearing here, in fact, the notion or principle of *gnōthi seauton* (an explicit reference to the Delphic principle).[71] However, it is interesting to see that this *gnōthi seauton*, appearing before any notion of care of the self, is given in a weak form. It is simply a counsel of prudence. It does not appear with the strong meaning it will have later. Socrates asks Alcibiades to reflect on himself a little, to review his life and compare himself with his rivals. A counsel of prudence: Think a bit about who you are in comparison with those you want to confront and you will discover your inferiority.

His inferiority consists in this: You are not only not wealthy and have not received any education, but also you cannot compensate for these defects (of wealth and education) by the only thing which would enable you to confront them without too much inferiority—a know-how (*savoir*), a *tekhnē*.[72] You do not have the *tekhnē* that would enable you to compensate for these initial inferiorities. Here Socrates demonstrates to Alcibiades that he does not have the *tekhnē* to enable him to govern the city-state well and be at least on an equal footing with his rivals. Socrates demonstrates this to him through a process which is absolutely classical in all the Socratic dialogues: What is it to govern the city well; in what does good government of the city consist; how do we recognize it? There is a long series of questions. We end up with this definition advanced by Alcibiades: The city is well governed when harmony reigns amongst its citizens.[73] Alcibiades is asked: What is this harmony; in what does it consist? Alcibiades cannot answer. The poor boy cannot answer and then despairs. He says: "I no longer know what I am saying. Truly, it may well be that I have lived for a long time in a state of shameful ignorance without even being aware of it."[74] To this Socrates responds: Don't worry; if you were to discover your shameful ignorance and that you do not even know what you are saying when you are fifty, it really would be difficult for you to remedy it, because it would be very difficult to take care of yourself (to take pains with oneself: *epimelēthēnai sautou*). However, "here you are at the time of life when one ought to be aware of it."[75] I would like to stop for a moment on this first appearance in philosophical discourse— subject once again to the dating of the *Alcibiades*—of this formula "taking care of oneself," "taking pains with oneself."

First, as you can see, the need to be concerned about the self is linked to the exercise of power. We have already come across this in the Lacedaemonian or Spartan maxim of Anaxandridas. Except, however, that in the apparently traditional formula—"We entrust our lands to our helots so that we can take care of ourselves"—"taking care of oneself' was the consequence of a statutory situation of power. Here, rather, you see that the question of the care of oneself, the theme of the care of oneself, does not appear as an aspect of statutory privilege. It appears rather as a condition for Alcibiades to pass from his position of statutory privilege (grand, rich, traditional family, etcetera) to definite political action, to actual government of the city-state. However, you can see that "taking care of oneself" is entailed by and inferred from the individual's will to exercise political power over others. One cannot govern others, one cannot govern others well, one cannot transform one's privileges into political action on others, into rational action, if one is not concerned about oneself. Care of the self: the point at which the notion emerges is here, between privilege and political action.

Second, you can see that this notion of care of the self, this need to be concerned about oneself, is linked to the inadequacy of Alcibiades' education. But the target here is, of course, Athenian education itself, which is wholly inadequate in two respects. It is inadequate in its specifically pedagogical aspect (Alcibiades' master was worthless, a slave, and an ignorant slave, and the education of a young aristocrat destined for political career is too important to be handed over to a family slave).

There is also criticism of the other aspect, which is less immediately clear but lurks throughout the beginning of the dialogue: the criticism of love, of the *erōs* of boys, which has not had the function for Alcibiades it should have had, since Alcibiades has been pursued by men who really only want his body, who do not want to take care of him—the theme reappears a bit later—and who therefore do not encourage Alcibiades to take care of himself. Furthermore, the best proof of their lack of interest in Alcibiades himself, of their lack of concern that he should be concerned about himself, is that they abandon him to do what he wants as soon as he loses his desirable youth. The need for the care of the self is thus inscribed not only within the political project, but also within the pedagogical lack.

Third, something as important as and immediately connected to the former feature is the idea that it would be too late to rectify matters if Alcibiades were fifty. This was not the age for taking care of oneself. One must learn to take care of oneself at the critical age when one leaves the hands of the pedagogues and enters political activity. To a certain extent, this text contradicts or raises a problem with regard to another text I read to you a short while ago, the *Apology*, in which Socrates, defending himself in front of his judges, says: But the job I have followed in Athens was an important one. It was entrusted to me by the gods and consisted in placing myself in the street and stopping everyone, young and old, citizens and noncitizens, to tell them to take care of themselves.[76] Here, the *epimeleia heautou* appears as a general function of the whole of life, whereas in the *Alcibiades* it appears as a necessary moment of the young man's training. A very important question, a major debate and a turning point in the care of the self, arises when the care of the self in Epicurean and Stoic philosophy becomes a permanent obligation for every individual throughout his life. But in this, if you like, early Socratic-Platonic form, the care of the self is, rather, an activity, a necessity for young people, within a relationship between them and their master, or them and their lover, or them and their master and lover. This is the third point, the third characteristic of the care of the self. Fourth, and finally, the need to take care of the self does not appear to be urgent when Alcibiades formulates his political projects, but only when he sees that he is unaware of . . . what? Well, that he is unaware of the object itself, of the nature of the object he has to take care of. He knows that he wants to take care of the city-state. His status justifies him doing this. But he does not know how to take care of the city-state; he does not know in what the purpose and end of his political activity will consist (the well-being of the citizens, their mutual harmony). He does not know the object of good government, and that is why he must pay attention to himself.

So, two questions arise at this point, two questions to be resolved that are directly linked to each other. We must take care of the self. But this raises the question: What, then, is this self with which we must be concerned when we are told that we must care about the self? I refer you to the passage that I will comment upon at greater length next time, but which is very important. The dialogue of *Alcibiades* has a subtitle, but one which was added much later, in the Alexandrian period I think, but I am not sure and will have to check for next time. This subtitle is "of human nature."[77] Now when you consider the development of the whole last part of the text—which begins at the passage I pointed out to you—you see that the question Socrates poses and attempts to resolve is not: You must take care of yourself now you are a man, and so I ask, what is a man? Socrates asks a much more precise, interesting, and difficult question, which is: You must take care of yourself; but what is this "oneself" (*auto to auto*)[78] since it is your self you must take care of? Consequently the question does not concern the nature of man but what we—that is us today, since the word is not in the Greek text—will call the question of the subject. What is this subject, what is this point towards which this reflexive activity, this reflected activity, which turns the individual back to himself, must be directed? The first question, then, is what is this self?

The second question to be resolved is: If we develop this care of the self properly, if we take it seriously, how will it be able to lead us, and how will it lead Alcibiades to what he wants, that is to say to knowledge of the *tekhnē* he needs to be able to govern others, the art that will enable him to govern well? In short, what is at stake in the whole of the second part, of the end of the dialogue, is this: "oneself," in the expression "caring about oneself," must be given a definition which entails, opens up, or gives access to a knowledge necessary for good government. What is at stake in the dialogue, then, is this: What is this self I must take care of in order to be able to take care of the others I must govern properly? This circle, [which goes] from the self as an object of care to knowledge of government as the government of others, is, I think, at the heart of the end of this dialogue. Anyway, the question of "caring about oneself" first emerges in ancient philosophy on the back of this question. So, thank you, and next week we will begin again at 9:15. I will try to conclude this reading of the dialogue.

The Courage of the Truth

Michel Foucault

Michel Foucault (1926–1984) was a philosopher and historian and one of the most original and dynamic thinkers of the postwar period. Foucault's lecture course, delivered in 1983–1984 at the Collège de France, continues his concern for thinking through the relations of ethics, care, and truth in Greek and Roman philosophy. The 1984 course gave particular attention to *parrhēsia*—frankness of speech—a term and theme that he had explored in the previous year's lectures. *Parrhēsia*, as Foucault explains, is not only a mode of speech but is also an essential part of an ethical form of life to the degree that it forms part of the way in which we govern ourselves and others, seek the truth, and undertake practices of care.

This year I would like continue with the theme of *parrhēsia*, truth-telling, that I began to talk about last year. The lectures I would like to give will no doubt be somewhat disjointed because they deal with things that I would like to have done with, as it were, in order to return, after this several-years-long Greco-Latin "trip,"[1] to some contemporary problems which I will deal with either in the second part of the course, or possibly in the form of a working seminar.

Well then, I shall remind you of something. You know that the rules are that the lectures of the Collège are and must be public. So it is quite right that anyone, French citizens or otherwise, has the right to come and listen to them. The Collège professors are obliged to report regularly on their research in these public lectures. However, this principle poses problems and raises a number of difficulties, because the work, the research one may undertake—especially [with regard to] questions like those I dealt with previously [and] to which I would now like to return, that is to say the analysis of certain practices and institutions in modern society—increasingly involves collective work which, of course, can only be pursued in the form of a closed seminar, and not in a room like this and with such a large public.[2] I am not going to hide from you the fact that I shall raise the problem of whether it is possible, whether it may be institutionally acceptable to divide the work I am doing here between public lectures—which, once again, are part of the job and of your rights—and lectures which would be restricted to small working groups with some students or researchers who have a more specialized interest in the question being studied. The public lectures would be, as it were, the exoteric version of the somewhat more esoteric work in a group. In any case, I don't know how many public lectures I will give or for how long. So, if you like, let's get going and then we'll see.

This year I would like to continue the study of free-spokenness (*franc-parler*), of *parrhēsia* as modality of truth-telling. I will restate the general idea for those of you who were not here last year. It is absolutely true that the analysis of the specific structures of those discourses which claim to be and are accepted as true discourse is both interesting and important. Broadly speaking, we could call the analysis of these structures an epistemological analysis. On the other hand, it seemed to me that it would be equally interesting to analyze the conditions and forms of the type of act by which the subject *manifests* himself when speaking the truth, by which I mean, thinks of himself and is recognized by others as speaking the truth. Rather than analyzing the forms by which a discourse is recognized as true, this would involve analyzing the form in which, in his act of telling the truth, the individual constitutes himself and is constituted by others as a subject of a discourse of truth, the form in which he presents himself to himself and to others as someone who tells the truth, the form of the subject telling the truth. In contrast with the study of epistemological structures, the analysis of this domain could be called the study of "alethurgic" forms. I am using here a word which I commented on last year or two years ago. Etymologically, alethurgy would be the production of truth, the act by which truth is manifested.[3] So, let's leave the kind of analysis which focuses on "epistemological structure" to one side and begin to analyze "alethurgic forms." This is the framework in which I am studying the notion and practice of *parrhēsia*, but for those of you who were not here I would like to recall how I arrived at this problem. I came to it from the old, traditional question, which is at the very heart of Western philosophy, of the relations between subject and truth, a question which I posed, which I took up first of all in classical, usual, and traditional terms, that is to say: on the basis of what practices and through what types of discourse have we tried to tell the truth about the subject? Thus: on the basis of what practices, through what types of discourse have we tried to tell the truth about the mad subject or the delinquent subject?[4] On the basis of what discursive practices was the speaking, laboring, and living subject constituted as a possible object of knowledge (*savoir*)?[5]

This was the field of study that I tried to cover for a period. And then I tried to envisage this same question of subject/truth relations in another form: not that of the discourse of truth in which the truth about the subject can be told, but that of the discourse of truth which the subject is likely and able to speak about himself, which may be, for example, avowal, confession, or examination of conscience. This was the analysis of the subject's true discourse about himself, and it was easy to see the importance of this discourse for penal practices or in the domain of the experience of sexuality.[6]

This theme, this problem led me, in previous years' lectures, to [attempt] the historical analysis of practices of telling the truth about oneself. In undertaking this analysis I noticed something completely unexpected. To be more precise, I shall say that it is easy to note the great importance of the principle that one should tell the truth about oneself in all of ancient morality and in Greek and Roman culture. In support and as illustration of the importance of this principle in ancient culture, we can cite such frequently, constantly, continually recommended practices [as] the examination of conscience prescribed by the Pythagoreans or Stoics, of which Seneca provides such elaborate examples, and which are found again in Marcus Aurelius.[7] We can also cite practices like correspondence, the exchange of moral, spiritual letters, examples of which can be found in Seneca, Pliny the Younger, Fronto, and Marcus Aurelius.[8] We can also cite, again as illustration of this principle "one should tell the truth about oneself," other, perhaps less well-known practices which have left fewer traces, like the notebooks, the kinds of journals which people were recommended to keep about themselves, either for the recollection and meditation of things one has experienced or read, or to record one's dreams when waking up.[9]

So it is quite easy to locate a very clear and solid set of practices in ancient culture which involve telling the truth about oneself. These practices are certainly not unknown and I make no claim to having discovered them; that is not my intention. But I think there is a consistent tendency to analyze these forms of practices of telling the truth about oneself by relating them, as it were, to a central axis which is, of course—and entirely legitimately— the Socratic principle of "know yourself": they are then seen as the illustration, the implementation, the concrete exemplification of the principle of *gnōthi seauton*. But I think it would be interesting to situate these practices in a broader context defined by a principle of which the *gnōthi seauton* is itself only an implication. This principle—I think I tried to bring this out in the lectures I gave two years ago—is that of *epimeleia heautou* (care of self, application to oneself).[10] This precept, which is so archaic, so ancient in Greek and Roman culture, and which in Platonic texts, and [more] precisely in the Socratic dialogues, is regularly associated with the *gnōthi seauton*, this principle (*epimelē seautō*: take care of yourself) gave rise, I think, to the development of what could be called a "culture of self"[11] in which a whole set of practices of self are formulated, developed, worked out, and transmitted. Studying these practices of self as the historical framework in which the injunction "one should tell the truth about oneself" developed, I saw a figure emerge who was constantly present as the indispensable partner, at any rate the almost necessary helper in this obligation to tell the truth about oneself. To put it more clearly and concretely, I shall say: we do not have to wait until Christianity, until the institutionalization of the confession at the start of the thirteenth century,[12] until the organization and installation of a pastoral power,[13] for the practice of telling the truth about oneself to rely upon and appeal to the presence of the other person who listens and enjoins one to speak, and who speaks himself. In ancient culture, and therefore well before Christianity, telling the truth about oneself was an activity involving several people, an activity with other people, and even more precisely an activity with one other person, a practice for two. And it was this other person who is present, and necessarily present in the practice of telling the truth about oneself, which caught and held my attention.

The status and presence of this other person who is so necessary for me to be able to tell the truth about myself obviously poses some problems. It is not so easy to analyze, for if it is true that we are relatively familiar with the other who is necessary for telling the truth about oneself in Christian culture, in which he takes the institutional form of the confessor or spiritual director, and if it is fairly easy to spot this other person in modern culture, whose status and functions should no doubt be analyzed more precisely— this other person who is indispensable for me to be able to tell the truth about myself, whether in the role of doctor, psychiatrist, psychologist, or psychoanalyst—on the other hand, in ancient culture, where this role is nevertheless well attested, we have to acknowledge that its status is much more variable, vague, much less clear cut and institutionalized. In ancient culture this other who is necessary for me to be able to tell the truth about myself might be a professional philosopher, but he could be anybody. You recall, for example, the passage in Galen on the cure of errors and passions, in which he says that to tell the truth about oneself and to know oneself we need someone else whom we can pick up almost anywhere, so long as he is old enough and serious.[14] This person may be a professional philosopher, or he may be just anybody. He may be a teacher who is more or less part of an institutionalized pedagogical structure (Epictetus directed a school),[15] but he may be a personal friend, or a lover. He may be a provisional guide for a young man who is not yet fully mature, who has not yet made his basic choices in life, who is not yet the full master of himself, but he may also be a permanent adviser who will accompany someone throughout his life and guide him until death. You recall, for example, the Cynic Demetrius who was the counselor of Thrasea Paetus, an important figure in Roman political life in the middle of the first century, and who served him as counselor until the day of his death, until his suicide—since Demetrius was present at the suicide of Thrasea Paetus and conversed with him until his last breath about the immortality of the soul, naturally in the manner of the Socratic dialogue.[16]

The status of this other person is variable therefore. Nor is it any easier to isolate and define his role, his practice, since in one respect it is connected with and leans on pedagogy, but it is also guidance of the soul. It may also be a sort of political advice. But equally the role may be presented metaphorically and even manifest itself and take shape as a sort medical practice, since it is a question of taking care of the soul[17] and of fixing a regimen of life, which includes, of course, the regimen of passions, but also the dietary regimen,[18] and the mode of life in all its aspects.

However, even if the role of this other person who is indispensable for telling the truth about oneself is uncertain or, if you like, polyvalent, even if it appears with a number of different aspects and profiles—medical, political, and pedagogical—which mean that it is not always easy to grasp exactly what his role is, even so, whatever his role, status, function, and profile may be, this other has, or rather should have a particular kind of qualification in order to be the real and effective partner of truth-telling about self. And this qualification, unlike the confessor's or spiritual director's in Christian culture, is not given by an institution and does not refer to the possession and exercise of specific spiritual powers. Nor is it, as in modern culture, an institutional qualification guaranteeing a psychological, psychiatric, or psychoanalytic knowledge. The qualification required by this uncertain, rather vague, and variable character is a practice, a certain way of speaking which is called, precisely, *parrhēsia* (free-spokenness).

To be sure, it has now become quite difficult for us to recapture this notion of *parrhēsia*, of speaking out freely, constitutive of the figure of this other person who is indispensable for me to be able to tell the truth about myself. But it has nonetheless left many traces in the Latin and Greek texts. In the first place, it has obviously left traces in the fairly frequent use of the word, and then also through references to the notion even when the word itself is not used. We find many examples, in Seneca in particular, where the practice of *parrhēsia* is very clearly picked out in descriptions and characterizations, practically without the word being used, if only because of the difficulties the Latins had translating the word *parrhēsia* itself.[19] Apart from these occurrences of the word or references to the notion, there are also some texts which are more or less wholly devoted to the notion of *parrhēsia*. From the first century before Jesus Christ, there is the text of the Epicurean Philodemus, who wrote a *Peri parrhēsia*, a large part of which is sadly lost.[20] But there is also Plutarch's treatise, *How to Distinguish the Flatterer from the Friend*, which is entirely taken up with an analysis of *parrhēsia*, or rather of the two opposed, conflicting practices of flattery, on the one hand, and *parrhēsia* (free-spokenness) on the other.[21] There is Galen's text, which I referred to a moment ago, on the cure of errors and passions, in which a whole section is devoted to *parrhēsia* and to the choice of the person who is rightly qualified as being able and having to use this free-spokenness so that the individual can, in turn, tell the truth about himself and constitute himself as subject telling the truth about himself.[22] So this is how I was led to focus on this notion of *parrhēsia* as a constitutive component of truth-telling about self or, more precisely, as the element which qualifies the other person who is necessary in the game and obligation of speaking the truth about self.

You may recall that last year I undertook the analysis of this free-spokenness, of the practice of *parrhēsia*, and of the character able to employ *parrhēsia*, who is called the *parrhēsiast* (*parrhēsiastēs*)—the word appears later. The study of *parrhēsia* and of the *parrhēsiastēs* in the culture of self in Antiquity is obviously a sort of prehistory of those practices which are organized and developed later around some famous couples: the penitent and the confessor, the person being guided and the spiritual director, the sick person and the psychiatrist, the patient and the psychoanalyst. It was, in a sense, this prehistory that I was trying to write.

Only then, while studying this parrhesiastic practice in this perspective, as the prehistory of these famous couples, I became aware again of something which rather surprised me and which I had not foreseen. Although *parrhēsia* is an important notion in the domain of spiritual direction, spiritual guidance, or soul counseling, and however important it may be in Hellenistic and Roman literature in particular, it is important to recognize that its origin lies elsewhere, that it is not essentially, fundamentally, or primarily in the practice of spiritual guidance that it emerges.

Last year I tried to show you that the notion of *parrhēsia* was first of all and fundamentally a political notion. And this analysis of *parrhēsia* as a political notion, as a political concept, clearly took me away somewhat from my immediate project: the ancient history of practices of telling the truth about oneself. However, on the other hand, this drawback was compensated for by the fact that by taking up again or undertaking the analysis of *parrhēsia* in the field of political practices, I drew a bit closer to a theme which, after all, has always been present in my analysis of the relations between the subject and truth: that of relations of power and their role in the interplay between the subject and truth.

With the notion of *parrhēsia*, originally rooted in political practice and the problematization of democracy, then later diverging towards the sphere of personal ethics and the formation of the moral subject,[23] with this notion with political roots and its divergence into morality, we have, to put things very schematically—and this is what interested me, why I stopped to look at this and am still focusing on it—the possibility of posing the question of the subject and truth from the point of view of the practice of what could be called the government of oneself and others. And thus we come back to the theme of government which I studied some years ago.[24] It seems to me that by examining the notion of *parrhēsia* we can see how the analysis of modes of veridiction, the study of techniques of governmentality, and the identification of forms of practice of self interweave. Connecting together modes of veridiction, techniques of governmentality, and practices of the self is basically what I have always been trying to do.[25]

And to the extent that this involves the analysis of relations between modes of veridiction, techniques of governmentality, and forms of practice of self, you can see that to depict this kind of research as an attempt to reduce knowledge (*savoir*) to power, to make it the mask of power in structures, where there is no place for a subject, is purely and simply a caricature. What is involved, rather, is the analysis of complex relations between three distinct elements none of which can be reduced to or absorbed by the others, but whose relations are constitutive of each other. These three elements are: forms of knowledge (*savoirs*), studied in terms of their specific modes of veridiction; relations of power, not studied as an emanation of a substantial and invasive power, but in the procedures by which people's conduct is governed; and finally the modes of formation of the subject through practices of self. It seems to me that by carrying out this triple theoretical shift—from the theme of acquired knowledge to that of veridiction, from the theme of domination to that of governmentality, and from the theme of the individual to that of the practices of self—we can study the relations between truth, power, and subject without ever reducing each of them to the others.[26]

Now, having recalled this general trajectory, I would like [to mention] briefly some of the essential elements which characterize *parrhēsia* and the parrhesiastic role. Very briefly, for a few minutes, and once again [for the benefit of] those who were not here, I shall go back over some things I have already said (I apologize to those who will be hearing this again), and then I would like, as quickly as possible, to move on to another way of envisaging the same notion of *parrhēsia*. You recall that, etymologically, *parrhēsia* is the activity that consists in saying everything: *pan rēma*. *Parrhēsiazesthai* is "telling all." The *parrhēsiastēs* is the person who says everything.[27] Thus, as an example, in his discourse *On the Embassy*, Demosthenes says: It is necessary

to speak with *parrhēsia*, without holding back at anything, without concealing anything.[28] Similarly, in the *First Philippic* he takes up exactly the same
term and says: I will tell you what I think without concealing anything.[29]
The parrhesiast is the person who tells all.

But we should immediately add the clarification that this word *parrhēsia*
may be employed with two values. I think we find it used in a pejorative
sense, first in Aristophanes, and afterwards very commonly, even in Christian literature. Used in a pejorative sense, *parrhēsia* does indeed consist in
saying everything, but in the sense of saying anything (anything that comes
to mind, anything that serves the cause one is defending, anything that
serves the passion or interest driving the person who is speaking). The parrhesiast then becomes and appears as the impenitent chatterbox, someone
who cannot restrain himself or, at any rate, someone who cannot index-link
his discourse to a principle of rationality and truth. There is an example of
this use of the term *parrhēsia* in a pejorative sense (saying everything, saying
anything, saying whatever comes to mind without reference to any principle
of reason or truth) in Isocrates, in the discourse entitled *Busiris*, in which
Isocrates says that, unlike the poets who ascribe everything and anything,
absolutely every and any qualities and defects to the gods, one should not
say everything about them.[30] Similarly, in Book VIII of *The Republic* (I will
give you the exact reference shortly because I will come back to this text)
there is the description of the bad democratic city, which is all motley, frag-
mented, and dispersed between different interests, passions, and individuals who do not agree with each other. This bad democratic city practices
parrhēsia: anyone can say anything.[31]

But the word *parrhēsia* is also employed in a positive sense, and then
parrhēsia consists in telling the truth without concealment, reserve, empty
manner of speech, or rhetorical ornament which might encode or hide it.
"Telling all" is then: telling the truth without hiding any part of it, without
hiding it behind anything. In the *Second Philippic*, Demosthenes thus says
that, unlike bad parrhesiasts who say anything and do not index their discourses to reason, he, Demosthenes, does not want to speak without reason,
he does not want to "resort to insults" and "exchange blow for blow"[32] (you
know, those infamous disputes in which anything is said so long as it may
harm the adversary and be useful to one's own cause). He does not want to
do this, but rather he wants to tell the truth (*ta alethē*: things that are true)

with *parrhēsia* (*meta parrhēsias*). Moreover, he adds: I will conceal nothing (*oukh apokhrupsōmai*).[33] To hide nothing and say what is true is to practice *parrhēsia*. *Parrhēsia* is therefore "telling all," but tied to the truth: telling the whole truth, hiding nothing of the truth, telling the truth without hiding it behind anything.

However, I don't think this suffices as a description and definition of this notion of *parrhēsia*. In fact—leaving aside the negative senses of the term for the moment—in addition to the rule of telling all and the rule of truth, two supplementary conditions are required for us to be able to speak of *parrhēsia* in the positive sense of the term. Not only must this truth really be the personal opinion of the person who is speaking, but he must say it as being what he thinks, [and not] reluctantly—and this is what makes him a parrhesiast. The parrhesiast gives his opinion, he says what he thinks, he personally signs, as it were, the truth he states, he binds himself to this truth, and he is consequently bound to it and by it. But this is not enough. For after all, a teacher, a grammarian or a geometer, may say something true about the grammar or geometry they teach, a truth which they believe, which they think. And yet we will not call this *parrhēsia*. We will not say that the geometer and grammarian are parrhesiasts when they teach truths which they believe. For there to be *parrhēsia*, you recall—I stressed this last year—the subject must be taking some kind of risk [in speaking] this truth which he signs as his opinion, his thought, his belief, a risk which concerns his relationship with the person to whom he is speaking. For there to be *parrhēsia*, in speaking the truth one must open up, establish, and confront the risk of offending the other person, of irritating him, of making him angry and provoking him to conduct which may even be extremely violent. So it is the truth subject to risk of violence. For example, in the *First Philippic*, after having said that he is speaking *meta parrhēsias* (with frankness), Demosthenes [adds]: I am well aware that, by employing this frankness, I do not know what the consequences will be for me of the things I have just said.[34]

In short, *parrhēsia*, the act of truth, requires: first, the manifestation of a fundamental bond between the truth spoken and the thought of the person who spoke it; [second], a challenge to the bond between the two interlocutors (the person who speaks the truth and the person to whom this truth is addressed). Hence this new feature of *parrhēsia*: it involves some form of courage, the minimal form of which consists in the parrhesiast taking the risk of breaking and ending the relationship to the other person which was precisely what made his discourse possible. In a way, the parrhesiast always risks undermining that relationship which is the condition of possibility of his discourse. This is very clear in *parrhēsia* as spiritual guidance, for example, which can only exist if there is friendship, and where the employment of truth in this spiritual guidance is precisely in danger of bringing into question and breaking the relationship of friendship which made this discourse of truth possible. But in some cases this courage may also take a maximal form when one has to accept that, if one is to tell the truth, not only may one's personal, friendly relationship with the person to whom one is speaking be brought into question, but one may even be risking one's life. When Plato goes to see Dionysius the Elder—this is recounted in Plutarch—he tells him truths which so offend the tyrant that he conceives the plan, which in fact he does not put into execution, of killing Plato. But Plato fundamentally knew and accepted this risk.[35] *Parrhēsia* therefore not only puts the relationship between the person who speaks and the person to whom he addresses the truth at risk, but it may go so far as to put the very life of the person who speaks at risk, at least if his interlocutor has power over him and cannot bear being told the truth. In the *Nicomachean Ethics*, Aristotle lays stress on the connection between *parrhēsia* and courage when he links what he calls *megalopsukhia* (greatness of soul) to the practice of *parrhēsia*.[36]

Only—and this is the last feature I would like to recall briefly—*parrhēsia* may be organized, developed, and stabilized in what could be called a parrhesiastic game. For if the parrhesiast is someone who, by telling the truth, the whole truth, regardless of any other consideration, risks bringing his relationship to the other into question, and even risks his life, on the other hand, the person to whom this truth is told—whether this is the assembled people deliberating on the best decisions to take, or the Prince, the tyrant or king to whom advice must be given, or the friend one is guiding—this person (people, king, friend), if he wants to play the role proposed to him by the parrhesiast in telling him the truth, must accept the truth, however much it may hurt generally accepted opinion in the Assembly, the Prince's passions or interests, or the individual's ignorance or blindness. The people, the Prince, and the individual must accept the game of *parrhēsia*; they must play it themselves and recognize that they have to listen to the person who takes the risk of telling them the truth. Thus the true game of *parrhēsia* will be established on the basis of this kind of pact which means that if the parrhesiast demonstrates his courage by telling the truth despite and regardless of everything, the person to whom this *parrhēsia* is addressed will have to demonstrate his greatness of soul by accepting being told the truth. This kind of pact, between the person who takes the risk of telling the truth and the person who agrees to listen to it, is at the heart of what could be called the parrhesiastic game.

So, in two words, *parrhēsia* is the courage of the truth in the person who speaks and who, regardless of everything, takes the risk of telling the whole truth that he thinks, but it is also the interlocutor's courage in agreeing to accept the hurtful truth that he hears.

You can see then how the practice of *parrhēsia* is opposed to the art of rhetoric in every respect. Very schematically, we can say that rhetoric, as it was defined and practiced in Antiquity, is basically a technique concerning the way that things are said, but does not in any way determine the relations between the person who speaks and what he says. Rhetoric is an art, a technique, a set of processes which enable the person speaking to say something which may not be what he thinks at all, but whose effect will be to produce convictions, induce certain conducts, or instill certain beliefs in the person [to whom he speaks]. In other words, rhetoric does not involve any bond of belief between the person speaking and what he [states]. The good rhetorician, the good rhetor is the man who may well say, and who is perfectly capable of saying, something completely different from what he knows, believes, and thinks, but of saying it in such a way that, in the final analysis, what he says—which is not what he believes, thinks, or knows—becomes what those he has spoken to think, believe, and think they know. The connection between the person speaking and what he says is broken in rhetoric, but the effect of rhetoric is to establish a constraining bond between what is said and the person or persons to whom it is said. You can see that from this point of view rhetoric is the exact opposite of *parrhēsia*, [which entails on the contrary a] strong, manifest, evident foundation between the person speaking and what he says, since he must openly express his thought, and you can see that in *parrhēsia* there is no question of saying anything other than what one thinks. *Parrhēsia* therefore establishes a strong, necessary, and constitutive bond between the person speaking and what he says, but it exposes to risk the bond between the person speaking and the person to whom he speaks. For, after all, it is always possible that the person to whom one is speaking will not welcome what one says. He may take offence at what one says, he may reject it and even punish or take revenge on the person who has told him the truth. So rhetoric does not entail any bond between the person speaking and what is said, but aims to establish a constraining bond, a bond of power between what is said and the person to whom it is said. *Parrhēsia*, on the other hand, involves a strong and constitutive bond between the person speaking and what he says, and, through the effect of the truth, of the injuries of truth, it opens up the possibility of the bond between the person speaking and the person to whom he has spoken being broken. Let's say, very schematically, that the rhetorician is, or at any rate may well be an effective liar who constrains others. The parrhesiast, on the contrary, is the courageous teller of a truth by which he puts himself and his relationship with the other at risk.

These are all things which I spoke to you about last year. I would like now to move on a bit and note straightaway that we should not think of *parrhēsia* as a sort of well-defined technique in a counterbalancing and symmetrical relation to rhetoric. We should not think that in Antiquity, facing the rhetorician who was a professional, a technician, and facing rhetoric, which was a technique and required an apprenticeship, there was a parrhesiast and a *parrhēsia* which would also be [Michel Foucault is interrupted at this point by pop music from one of the cassette recorders. We hear a member of the audience rush to their machine. M.F.: "I think you are mistaken. It is at least Michael Jackson? Too bad."]

The parrhesiast is not a professional. And *parrhēsia* is after all something other than a technique or a skill, although it has technical aspects. *Parrhēsia* is not a skill; it is something which is harder to define. It is a stance, a way of being which is akin to a virtue, a mode of action. *Parrhēsia* involves ways of acting, means brought together with a view to an end, and in this respect it has, of course, something to do with technique, but it is also a role which is useful, valuable, and indispensable for the city and for individuals. *Parrhēsia* should be regarded as a modality of truth-telling, rather than [as a] technique [like] rhetoric. To arrive at a better definition we can contrast it with other basic modalities of truth-telling found in Antiquity, and which will no doubt be found, in displaced and different guises and forms, in other societies, as well as our own. Basing ourselves on the clear understandings which

Antiquity has left us about these things, we may define four basic modalities of truth-telling.

First, the truth-telling of prophecy. I will not try here to analyze what the prophets said (the structures, as it were, of what was said by prophets), but rather the way in which the prophet constitutes himself and is recognized by others as a subject speaking the truth. Evidently, the prophet, like the parrhesiast, is someone who tells the truth. But I think that what fundamentally characterizes the prophet's truth-telling, his veridiction, is that the prophet's posture is one of mediation. The prophet, by definition, does not speak in his own name. He speaks for another voice; his mouth serves as intermediary for a voice which speaks from elsewhere. The prophet, usually, transmits the word of God. The discourse he articulates and utters is not his own. He addresses a truth to men which comes from elsewhere. The prophet's position is intermediary in another sense in that he is between the present and the future. The second characteristic of the prophet's intermediary position is that he reveals what time conceals from humans, what no human gaze could see and no human ear could hear without him. Prophetic truth-telling is also intermediary in that, in one way of course, the prophet reveals, shows, or sheds light on what is hidden from men, but in another way, or rather at the same time, he does not reveal without being obscure, and he does not disclose without enveloping what he says in the form of the riddle. Hence prophecy basically never gives any univocal and clear prescription. It does not bluntly speak the pure, transparent truth. Even when the prophet says what is to be done, one still has to ask oneself whether one has really understood, whether one may not still be blind; one still has to question, hesitate, and interpret.

Now *parrhēsia* contrasts with these different characteristics of prophetic truth-telling in each of these precise respects. You can see then that the parrhesiast is the opposite of the prophet in that the prophet does not speak for himself, but in the name of someone else, and he articulates a voice which is not his own. In contrast, the parrhesiast, by definition, speaks in his own name. It is essential that he expresses his own opinion, thought, and conviction. He must put his name to his words; this is the price of his frankness. The prophet does not have to be frank, even when he tells the truth. Second, the parrhesiast does not foretell the future. Certainly, he reveals and discloses what people's blindness prevents them from seeing, but he does not unveil the future. He unveils what is. The parrhesiast does not help people somehow to step beyond some threshold in the ontological structure of the human being and of time which separates them from their future. He helps them in their blindness, but their blindness about what they are, about themselves, and so not the blindness due to an ontological structure, but due to some moral fault, distraction, or lack of discipline, the consequence of inattention, laxity, or weakness. It is in this interplay between human beings and their blindness due to inattention, complacency, weakness, and moral distraction that the parrhesiast performs his role, which, as you can see, is consequently a revelatory role very different from that of the prophet, who stands at the point where human finitude and the structure of time are conjoined. Third, the parrhesiast, again by definition, and unlike the prophet, does not speak in riddles. On the contrary, he says things as clearly and directly as possible, without any disguise or rhetorical embellishment, so that his words may immediately be given their prescriptive value. The parrhesiast leaves nothing to interpretation. Certainly, he leaves something to be done: he leaves the person he addresses with the tough task of having the courage to accept this truth, to recognize it, and to make it a principle of conduct. He leaves this moral task, but, unlike the prophet, he does not leave the difficult duty of interpretation.

Second, I think we can also contrast parrhesiastic truth-telling with another mode of truth-telling which was very important in Antiquity, doubtless even more important for ancient philosophy than prophetic truth-telling: the truth-telling of wisdom. As you know, the sage—and in this he is unlike the prophet we have just been talking about—speaks in his own name. And even if this wisdom may have been inspired by a god, or passed on to him by a tradition, by a more or less esoteric teaching, the sage is nevertheless present in what he says, present in his truth-telling. The wisdom he expresses really is his own wisdom. The sage manifests his mode of being wise in what he says and, to that extent, although he has a certain intermediary function between timeless, traditional wisdom and the person he addresses, unlike the prophet, he is not just a mouthpiece. He is himself wise, a sage, and his mode of being wise as his personal mode of being qualifies him as a sage, and qualifies him to speak the discourse of wisdom. To that extent, insofar as he is present in his wise discourse and manifests his mode of being wise in his wise discourse, he is much closer to the parrhesiast than to the prophet. But the sage—and this is what characterizes him, at least through some of the traits that we can find in the ancient literature—keeps his wisdom in a state of essential withdrawal, or at least reserve. Basically, the sage is wise in and for himself, and does not need to speak. He is not forced to speak, nothing obliges him to share his wisdom, to teach it, or demonstrate it. This accounts for what might be termed his structural silence. And if he speaks, it is only because he is appealed to by someone's questions, or by an urgent situation of the city. This also explains why his answers—and then in this respect he may well be like the prophet and often imitate and speak like him—may well be enigmatic and leave those he addresses ignorant or uncertain about what he has actually said. Another characteristic of the truth-telling of wisdom is that wisdom says what is, unlike prophecy where what is said is what will be. The sage says what is, that is to say, he tells of the being of the world and of things. And if this telling the truth of the being of the world and of things has prescriptive value, it is not [in] the form of advice linked to a conjuncture, but in the form of a general principle of conduct.

These characteristics of the sage can be read and rediscovered in the text in which Diogenes Laertius portrays Heraclitus; it is a late text, but one of the richest in various kinds of information. First, Heraclitus lived in an essential withdrawal. He lived in silence. And Diogenes Laertius recalls the moment at which and why the break took place between Heraclitus and the Ephesians. The Ephesians had exiled his friend, Hermodorus, precisely because he was wise and better than them. They said: We want "there to be no one among us who is better than us."[37] And if there is someone who is better than us, let him go and live elsewhere. The Ephesians could not bear the superiority of precisely someone who tells the truth. They drove out the parrhesiast. They drove out Hermodorus, who was obliged to leave, forced into the exile with which they punished the person capable of telling the truth. Heraclitus, for his part, responded with voluntary withdrawal. Since the Ephesians have punished the best among them with exile, well, he says, all the others, who are less worthy, should be put to death. And since they are not put to death, I will be the one to leave. And from that time on, when asked to give laws to the city, he refused. Because, he says, the city is already dominated by a *ponēra politeia* (a bad mode of political life). So he withdraws himself and—in a famous image—plays knucklebones with children. To those who are indignant at him playing knucklebones with

children, he replies: "Why are you surprised, rascals, isn't this more worthwhile than administering the republic with you [*met'humōn politeuesthai*:
than conducting political life with you; M.F.]?"[38] He retires to the moun-
tains, practicing contempt of men (*misanthrōpon*).[39] And when asked why he
remained silent, he replied: "I keep quiet so that you may chatter."[40] Diogenes Laertius relates that in this retirement Heraclitus wrote his Poem in
deliberately obscure terms so that only those who were capable could read
it and so that he, Heraclitus, could not be despised for being read by all and
sundry.[41]

The figure and characteristics of the parrhesiast stand in contrast with
this role, this characterization of the sage, who basically remains silent, only
speaks when he really wants to, and [only] in riddles. The parrhesiast is not
someone who is fundamentally reserved. On the contrary, it is his duty,
obligation, responsibility, and task to speak, and he has no right to shirk this
task. We will see this precisely with Socrates, who recalls it frequently in
the *Apology*: the god has given him this office of stopping men, taking them
aside, and questioning them. And he will never abandon this office. Even
under the threat of death, he will carry out his task until the end, until his
final breath.[42] Whereas the sage keeps silent and responds only sparingly,
as little as possible, to the questions he may be asked, the parrhesiast is the
unlimited, permanent, unbearable questioner. Second, whereas the sage is
the person who, against the background of an essential silence, speaks in
riddles, the parrhesiast must speak, and he must speak as clearly as possible.
And finally, whereas the sage says what is, but in the form of the very being of things and of the world, the parrhesiast intervenes, says what is, but
in terms of the singularity of individuals, situations, and conjunctures. His
specific role is not to tell of the being of nature and things. In the analysis of
parrhēsia we will constantly find this opposition between useless knowledge
which speaks of the being of things and the world, on the one hand, and on
the other the parrhesiast's truth-telling which is always applied, questions,
and is directed to individuals and situations in order to say what they are
in reality, to tell individuals the truth of themselves hidden from their own
eyes, to reveal to them their present situation, their character, failings, the
value of their conduct, and the possible consequences of their decisions. The
parrhesiast does not reveal what is to his interlocutor; he discloses or helps
him to recognize what he is.

Finally, the third modality of truth-telling which can be contrasted with the parrhesiast's truth-telling is that of the professor, the technician, [the teacher]. The prophet, the sage, the person who teaches.

So, if you like, because maybe some of you are a bit weary from listening and others from not hearing, some from sitting down and others from standing, and me at any rate from speaking, we will stop for five or ten minutes. And then we will meet again shortly, OK? I will try to finish around 11.15. Thank you.

. . . In fact room 6 was not, is not, and will not be fitted with a public address system. You were told the truth when you were told that room 6 was not fitted with a public address system, but what you were not told, and what I was not told either, is that room 5 was. At any rate, it is now. So those of you who have had enough of standing up or sitting on the floor can find in room 5 a place where you will be able to sit down, read the newspaper, and chat peacefully. OK? There you are. So thanks and my apologies. So from now on, if I understand correctly, every Wednesday room 5 will be linked up with this room. It will no longer be rooms 8 and 6, but 8 and 5. That's it, my apologies for what has happened.

I have tried then to pick out the relationships and differences between the parrhesiastic mode of truth-telling and, first, the prophetic mode of truth-telling, and then that of wisdom. And now I would like to indicate, very schematically and allusively, some of the relations between parrhesiastic veridiction and the veridiction of someone who teaches—I would prefer to say, basically, of the technician. These characters (the doctor, the musician, the shoemaker, the carpenter, the teacher of armed combat, the gymnastics teacher), frequently mentioned by Plato in his Socratic and other dialogues, possess a knowledge characterized as *tekhnē*, know-how, that is to say, entailing particular items of knowledge, but taking shape in a practice and involving, for their apprenticeship, not only a theoretical knowledge, but a whole exercise (a whole *askēsis* or *meletē*).[43] They possess this knowledge, they profess it, and they are capable of teaching it to others. The technician, who possesses a *tekhnē*, has learned it, and is capable of teaching it, is someone obliged to speak the truth, or at any rate to formulate what he knows and pass it on to others; and, of course, this distinguishes him from the sage. After all, the technician has a certain duty to speak. He is obliged, in a way, to tell the knowledge he possesses and the truth he knows, because this knowledge and truth are linked to a whole weight of tradition. This man of *tekhnē* would not himself have been able to learn anything and today would know nothing at all, or very little, if there had not been, before him, a technician (*tekhnitēs*) like him, who had taught him, whose pupil he had been, and who had been his teacher. And just as he would not have learned anything if someone had not previously told him what they knew, so, in the same way, he will have to pass on his knowledge so that it does not die with him.

So, in this idea of someone with knowledge of *tekhnē*, someone who has received this knowledge and must pass it on, there is the principle of an obligation to speak which is not found in the sage but is found in the parrhesiast. But clearly, this teacher, this man of *tekhnē*, of expertise and teaching, does not take any risk in the truth-telling he has received and must pass on, and this is what distinguishes him from the parrhesiast. Everyone knows, and I know first of all, that you do not need courage to teach. On the contrary, the person who teaches establishes, or at any rate hopes or sometimes wants to establish a bond of shared knowledge, of heritage, of tradition, and possibly also of personal recognition or friendship, between himself and the person or persons who listen to him. Anyway, this truth-telling establishes a filiation in the domain of knowledge. Now we have seen that the parrhesiast, to the contrary, takes a risk. He risks the relationship he has with the person to whom he speaks. And in speaking the truth, far from establishing this positive bond of shared knowledge, heritage, filiation, gratitude, or friendship, he may instead provoke the other's anger, antagonize an enemy, he may arouse the hostility of the city, or, if he is speaking the truth to a bad and tyrannical sovereign, he may provoke vengeance and punishment. And he may go so far as to risk his life, since he may pay with his life for the truth he has told. Whereas, in the case of the technician's truth-telling, teaching ensures the survival of knowledge, the person who practices *parrhēsia* risks death. The technician's and teacher's truth-telling brings together and binds; the parrhesiast's truth-telling risks hostility, war, hatred, and death. And if the parrhesiast's truth may unite and reconcile, when it is accepted and the other person agrees to the pact and plays the game of *parrhēsia*, this is only after it has opened up an essential, fundamental, and structurally necessary moment of the possibility of hatred and a rupture.

We can say then, very schematically, that the parrhesiast is not the prophet who speaks the truth when he reveals fate enigmatically in the name of someone else. The parrhesiast is not a sage who, when he wants to and against the background of his silence, tells of being and nature (*phusis*) in the name of wisdom. The parrhesiast is not the professor or teacher, the expert who speaks of *tekhnē* in the name of a tradition. So he does not speak of fate, being, or *tekhnē*. Rather, inasmuch as he takes the risk of provoking war with others, rather than solidifying the traditional bond, like the teacher, by [speaking] in his own name and perfectly clearly, [unlike the] prophet who speaks in the name of someone else, [inasmuch as] finally [he tells] the truth of what is in the singular form of individuals and situations, and not the truth of being and the nature of things, the parrhesiast brings into play the true discourse of what the Greeks called *ēthos*.

Fate has a modality of veridiction which is found in prophecy. Being has a modality of veridiction found in the sage. *Tekhnē* has a modality of veridiction found in the technician, the professor, the teacher, the expert. And finally, *ēthos* has its veridiction in the speech of the parrhesiast and the game of *parrhēsia*. Prophecy, wisdom, teaching, and *parrhēsia* are, I think, four modes of veridiction which, [first], involve different personages, second, call for different modes of speech, and third, relate to different domains (fate, being, *tekhnē*, *ēthos*).

Actually, in this survey I am not essentially defining four historically distinct social types. I do not mean that there were four professions or four social types in ancient civilization: the prophet, the sage, the teacher, and the parrhesiast. Certainly, it may be that these four major modalities of truth-telling (prophetic, wise, technical, and ethical or parrhesiastic) correspond to quite distinct institutions, or practices, or personages. One of the reasons why the example of Antiquity is privileged is precisely that it enables us to separate out, as it were, these different [modalities] of truth-telling, these different modes of veridiction. Because, in Antiquity, they are fairly clearly distinguished and embodied, formulated, and almost institutionalized in different forms. There is the prophetic function, which was quite clearly defined and institutionalized. The character of the sage was also quite clearly picked out (see the portrait of Heraclitus). You see the teacher, the technician, the man of *tekhnē* appear very clearly in the Socratic dialogues (the Sophists were precisely these kinds of technicians and teachers who claimed to have a universal function). As for the parrhesiast, his specific profile appears very clearly—we will come back to this next week—with Socrates, and then with Diogenes and a series of other philosophers. However, as distinct as these roles may be, and even if at certain times, and in certain societies or civilizations, you see these four functions taken on, as it were, by very clearly distinct institutions or characters, it is important to note that fundamentally these are not social characters or roles. I insist on this; I would like to stress it: they are essentially modes of veridiction. It sometimes happens, and it will happen very often, even more often than not, that these modes of veridiction are combined with each other, and we find them in forms of discourse, types of institutions, and social characters which mix the modes of veridiction with each other.

Already you can see how Socrates puts together elements of prophecy, wisdom, teaching, and *parrhēsia*. Socrates is the parrhesiast.[44] But you recall: who gave him his function as parrhesiast, his mission to question people, to take them by the sleeve and tell them: Take some care of yourself? It was the Delphic god, the prophetic authority which returned this verdict. When asked who was the wisest man in Greece, it replied: Socrates. And it was in order to honor this prophecy, and also to honor the Delphic god laying down the principle of "know yourself," that Socrates undertook his mission.[45] His function as parrhesiast is not therefore unrelated to this prophetic function, from which he nevertheless maintains his distinctness. Equally, although a parrhesiast, Socrates has a relationship with wisdom. This is evident in several traits: his personal virtue, his self-control, his abstention from all pleasures, his endurance in the face of all kinds of suffering, and his ability to detach himself from the world. You recall the famous scene in which Socrates becomes insensible, remaining immobile, impervious to the cold when he was a soldier at war.[46] We should also not forget that Socrates has that, in a sense even more important feature of wisdom, which is a particular kind of silence, regardless of everything. Because Socrates does not speak, he does not deliver speeches, he does not say spontaneously what he knows. On the contrary, he claims to be someone who does not know, and who, not knowing and knowing only that he does not know, will remain reserved and silent, confining himself to questioning. Questioning is, if you like, a particular way of combining the essential reserve of the sage, who remains silent, with the duty of *parrhēsia* (that is to say, the duty to challenge and speak). Except that the sage remains silent because he knows and has the right not to speak of his knowledge, whereas Socrates remains silent by saying that he does not know, and by questioning everyone and anyone in the manner of the parrhesiast. So here again you can see that the parrhesiastic feature combines with the features of wisdom. And finally, of course, there is the relationship with the technician, the teacher. The Socratic problem is how to teach the virtue and knowledge required to live well or also to govern the city properly. You recall the *Alcibiades*.[47] You recall too—we will come back to this next week—the end of the *Laches*, where Socrates agrees to teach the sons of Lysimachus and [Melesias] to take care of themselves.[48] So Socrates is the parrhesiast, but, once again, with a permanent, essential relationship to prophetic veridiction, the veridiction of wisdom, and the technical veridiction of teaching.

So, prophecy, wisdom, teaching, technique, and *parrhēsia* should be seen much more as fundamental modes of truth-telling than as characters. There is the modality which speaks enigmatically about that which is hidden from every human being. There is the modality of truth-telling which speaks apodictically about being, *phusis*, and the order of things. There is the veridiction which speaks demonstratively about kinds of knowledge and expertise. There is finally the veridiction which speaks polemically about individuals and situations. These four modes of truth-telling are, I believe, absolutely fundamental for the analysis of discourse to the extent that, in discourse, the subject who tells the truth is constituted for himself and for others. I think that since Greek culture, the subject who tells the truth takes these four possible forms: he is either prophet, or sage, or technician, or parrhesiast. It would be interesting to investigate how these four modalities, which, again, once and for all, are not identified with roles or characters, are combined in different cultures, societies, or civilizations in different modes of discursivity, in what could be called the different "regimes of truth" found in different societies.

It seems to me—at any rate, this is what I have tried to show you, however schematically—that in Greek culture at the end of the fifth and the beginning of the fourth century BCE we can find these four major modes of veridiction distributed in a kind of rectangle: that of prophecy and fate, that of wisdom and being, that of teaching and *tekhnē*, and that of *parrhēsia* and *ēthos*. But if these four modalities are thus quite clearly decipherable, separable, and separated from each other at this time, one of the features of the history of ancient philosophy (and also no doubt of ancient culture generally) is that there is a tendency for the mode of truth-telling characteristic of wisdom and the mode of truth-telling characteristic of *parrhēsia* to come together, join together, to link up with each other in a sort of philosophical modality of truth-telling which is very different from prophetic truth-telling as well as from the teaching of *tekhnai*, of which rhetoric is an example. We will see a philosophical truth-telling separating off, or anyway the development of a philosophical truth-telling which will ever more insistently claim to speak of being or the nature of things only to the extent that this truth-telling concerns, is relevant for, is able to articulate and found a truth-telling about *ēthos* in the form of *parrhēsia*. And to that extent, we can say that, only up to a certain point, of course, wisdom and *parrhēsia* merge. Anyway, it is as though they are attracted to each other, that there is something like a phenomenon of gravitation of wisdom and *parrhēsia*, a gravitation which manifests itself in the famous characters of philosophers telling the truth of things, but above all telling their truth to men, throughout Hellenistic and Roman, or Greco-Roman culture. If you like, there is the possibility of an analysis of a history of the regime of truth concerning the relations between *parrhēsia* and wisdom.

If we take up again these four major fundamental modes I have been talking about, we could say that medieval Christianity produced other groupings. Greco-Roman philosophy brought together the modalities of *parrhēsia* and wisdom. It seems to me that in medieval Christianity we see another type of grouping bringing together the prophetic and parrhesiastic modalities. The two modalities of telling the truth about the future (about what is hidden from men by virtue of their finitude and the structure of time, about what awaits men and the imminence of the still hidden event), and then telling the truth to men about what they are, were brought together in a number of particular [types] of discourses, and also institutions. I am thinking of preaching and preachers, and especially of those preachers, starting with the Franciscans and Dominicans, who played an absolutely major role across the Western world and throughout the Middle Ages in the perpetuation, but also renewal and transformation [of] the experience of threat for the medieval world. These great preachers played the role of both prophet and parrhesiast in that society. Those who speak of the threatening imminence of the future, of the Kingdom of the Last Day, of the Final Judgment, or of approaching death, at the same time tell men what they are, and tell them frankly, with complete *parrhēsia*, what their faults and crimes are, and in what respects and how they must change their mode of being.

Counterposed to this, it seems to me that the same medieval society, the same medieval civilization tended to bring together the other two modes of veridiction: that of wisdom, which tells of the being of things and their nature, and that of teaching. Telling the truth of being and telling the truth of knowledge was the task of an institution which was as specific to the Middle Ages as was preaching: the University. Preaching and the University appear to me to be institutions specific to the Middle Ages, in which we see the functions I have spoken about grouping together, in pairs, and defining a regime of veridiction, a regime of truth-telling, which is very different from the regime we could find in the Hellenistic and Greco-Roman world, where instead it was *parrhēsia* and wisdom that were combined.

And what about the modern epoch, you may ask? I don't really know. It would no doubt have to be analyzed. We could say perhaps—but these are hypotheses, not even hypotheses: some almost incoherent remarks—that you find the prophetic modality of truth-telling in some political discourses, in revolutionary discourse. In modern society, revolutionary discourse, like all prophetic discourse, speaks in the name of someone else, speaks in order to tell of a future which, up to a point, already has the form of fate. The ontological modality of truth-telling, which speaks of the being of things, would no doubt be found in a certain modality of philosophical discourse. The technical modality of truth-telling is organized much more around science than teaching, or at any rate around a complex formed by scientific and research institutions and teaching institutions. And the parrhesiastic modality has, I believe, precisely disappeared as such, and we no longer find it except where it is grafted on or underpinned by one of these three modalities. Revolutionary discourse plays the role of parrhesiastic discourse when it takes the form of a critique of existing society. Philosophical discourse as analysis, as reflection on human finitude and criticism of everything which may exceed the limits of human finitude, whether in the realm of knowledge or the realm of morality, plays the role of *parrhēsia* to some extent. And when scientific discourse is deployed as criticism of prejudices, of existing forms of knowledge, of dominant institutions, of current ways of doing things—and it cannot avoid doing this, in its very development—it plays this parrhesiastic role. That's wanted I wanted to say to you.

I intended to begin to speak to you of *parrhēsia* as I want to study it this year. But what would be the point? I would have five minutes and then it would be necessary to start again next week. So, if you like, we will go for a coffee. I could tell you: I do want to reply to your questions, but I fear that it does not have much meaning in lecture theaters . . . [reply to a question from the public concerning the closed seminar:] I have two things to say to you, this question and then another little thing. Concerning the semi-nar, once again, there is an institutional and legal problem. In principle, we do not have the right to have a closed seminar. And when I had a closed seminar—the one on Pierre Rivière, for example, some of you may recall it—there were complaints. And in fact, legally, we don't have the right to have a closed seminar. Only, I think that for certain kinds of work there is a contradiction between, [on the one hand] asking professors to give a public account of their research, and then, [on the other] preventing them from having a closed seminar where they can undertake research with some students. In other words, a professor can be asked to report on his research in public lectures, and nothing other than this, if he is doing research that he can undertake on his own. And, if you like, one of the purely technical reasons why, in fact, for some years I have lectured on ancient philosophy, is that, after all, it suffices to have the two hundred volumes of Budé available, and there you are. You don't need a working group. But if—as I would like to do—I want to study the practices, forms, and rationalities of government in modern society, I can only really do this in a group.

Now you can well understand—it is not offending anyone here—that this audience won't be able to function as a team. So what I would like is the right to divide the teaching in two: a public teaching, which is statutory; but also a teaching or research in a closed group which is, I think, the condi-tion for being able to carry out, or in any case replenish the public teaching we give. There is, I think, a contradiction in asking people to undertake research and public teaching if they are not given the institutional supports which make the research they have to do possible. So second, a small thing, it is probable—you know that I never really know what I will be doing from one week to the next—that either next week or the following week I will give a lecture, [or] half a lecture, on one of Dumézil's last two books, the one, you know, on "the black monk in gray" which concerns Nostradamus and includes a second part on Socrates (the *Phaedo* and the *Crito*). So as it is a difficult text, if some of you wish or have the opportunity to read it before—obviously, there is no obligation, we are not in a closed seminar, you do as you like—I would very much like to talk about it, certainly in two weeks, or maybe next week.

[question from the public:]—In a seminar or in the lectures?—The lectures. It's just that I am well aware that if I want to give a lecture on this, it presupposes to some degree that people have an idea of what is in the book. That's it, many thanks.

Anthropos Today: Reflections on Modern Equipment

Paul Rabinow

Paul Rabinow is an American anthropologist and philosopher. He has consistently focused his work on modernity as an anthropological problem and on the shifting relations of knowledge, power, and ethics, which constitute that problem. In the essays presented here, Rabinow identifies the need for anthropologists to develop conceptual tools—"equipment"— calibrated to a core problematic: namely, that *anthrōpos*—(the) human (being)—is the kind of being that suffers from too many discourses and knowledge claims about itself. Rabinow's diagnosis of such heterogeneity in these essays provided the equipment for his subsequent exploration of a problem space, which he has called an anthropology of the contemporary.

> *Paraskeuē* [equipment] . . . is the medium through which logos is transformed into ethos.
>
> — MICHEL FOUCAULT[1]

This book is proposed as a meditation on Michel Foucault's claim that "equipment is the medium of transformation of logos into ethos." A good deal of work is required, however, to grasp what such a claim might mean. The difficulty in part lies in the fact that the terms "equipment" and "meditation" are used in a distinctive technical sense. Furthermore, why one would want to transform "logos" into "ethos" equally requires explanation. Hence the reader is alerted that reading this book will require a certain patience. Additionally, and unexpectedly, the book addresses the reader as a friend. Initially this appellation too is opaque. However, using as a guide Jean Paul's wonderful claim that "Philosophy is the ability to make friends through the medium of a written text," we at least have some sense of the territory to be visited in the following chapters, as well as the manner in which that territory is to be traversed.[2]

A central purpose of the book is to assemble a toolkit of concepts. The goal of such a toolkit is to advance inquiry. The currently reigning modes of research in the human sciences are, it seems to me, deficient in vital respects. Those deficiencies are especially marked in the strained relations between an ever-accumulating body of information, the ways that information is given narrative and conceptual form, and how this knowledge fits into a conduct of life. No doubt all of this demands further elaboration, and this book attempts to respond to that demand.

The term "interpretive analytics" was coined by Hubert Dreyfus and myself and put to use in our book *Michel Foucault: Beyond Structuralism and Hermeneutics*.[3] Although the term cannot be said to have gained any special currency in the human sciences, I still find it useful. We arrived at the term while attempting to make Foucault's method more precise and explicit. Our claim was that Foucault was trying to move beyond the two methodological poles then dominant in the human sciences: a version of structuralism in which human signifying practice is seen as generating object-like, rule-governed semiotic systems that produce subjects as a function of discourse; and various versions of hermeneutics that found subjects and cultures infused with deep meaning they themselves had spun, webs of signification requiring interpretation. Foucault, we wrote,

> sought to avoid the structuralist analysis which eliminates notions of meaning altogether and substitutes a formal model of human behavior as rule-governed transformations of meaningless elements; to avoid the phenomenological project of tracing all elements back to the meaning-giving activity of an autonomous, transcendental subject; and finally, to avoid the attempt of commentary to read off the implicit meanings of social practices as well as the hermeneutic unearthing of a different and deeper meaning of which social actors are only dimly aware.[4]

Foucault had pieced together an innovative method through his tacking between so-called archaeological and genealogical emphases. Foucault, we argued, had gotten beyond structuralism and hermeneutics by showing how the historical relations of knowledge and power had produced an object of knowledge that was also the subject of knowledge: Man. Further, we concluded that the strengths and weaknesses of Foucault's writings could not be evaluated or appreciated adequately in terms of a correspondence theory of truth any more than through a deconstructive dissipation of the real. Rather, it seemed clear that the power of his work rested on its heuristic value.

There is a lineage of major work in the twentieth-century human sciences that has succeeded in bringing philosophical learning, diagnostic rigor, and a practice of inquiry that operates in proximity to concrete situations into a productive relationship. Such inquiry proceeds through mediated experience. It contributes to what used to be called a *Bildung*, a process of self-formation, that today might be called an attitude or an ethos. The proximity to concreteness is both the goal and the means through which inquiry operates when it works well. Understanding is a conceptual, political, and ethical practice. It is conceptual because without concepts one would not know what to think about or where to look in the world. It is political because reflection is made possible by the social conditions that enable this practice (thought may be singular, but it is not individual). It is ethical because the question of why and how to think are questions of what is good in life. Finally, all action is stylized; hence it is aesthetic, insofar as it is shaped and presented to others.

The goal of the meditations that follow is neither to systematically survey any specific domain of knowledge nor to solve any particular contemporary dispute. Rather, this book seeks to bring together a set of conceptual tools and to use them as a starting point to advance an experimental mode for the human sciences in which concepts and techniques could be made to function differently. By differently, I mean better. By better, I mean in a more sagacious manner. By a more sagacious manner, I mean a wiser one: logos serving phronesis, phronesis under the sign of philosophy, philosophy under the sign of ethos. By ethos I refer to that space of practice at the interface of ethics and culture. It is a premise of this work that both of the latter terms are very much in question today.

Hetero-Logoi

How to think about things human is a problem. Most attempts to solve this problem deploy one or another answer that claims to offer generality and stability. These attempts have produced incompatible answers. The fact that there is a problem in thinking about human things, and that part of that problem lies in the inability to provide a stable solution, is coexistent and cotemporal with the practice itself. This state of affairs has existed from the beginnings of Western philosophy, continued through the disputatious elaboration of theology, through the proliferation of what came to be known as the natural and social sciences, and through the strife of critical theory in the twentieth century, and again today is blazing afresh among, amidst, and between different sciences. However, the form of the problem—and there-fore the practices that produce it and that it produces—has not always been the same. We can conclude with some confidence, in a pragmatic spirit, that future attempts to define what the "thinking," or the "problem," "really is" are themselves fated to fail, by which I mean they will not establish them-selves as enduring solutions. They will join the cacophony of dispute that is such a vexing aspect of the subject matter itself.

No consensus has ever been reached about principles, methods, and modes of problem specification, or about modes, methods, and principles of verification, or about forms of narration in the human sciences. The hope for a positive science, or the end of metaphysics, or hermeneutical closure on the Bible or other authoritative texts, is like a cargo cult, which persists in the face of constant disappointment. How can thinkers fail to notice that almost no one outside their own immediate circle is paying attention to their proofs, their prophecies, their purges? When attempts have been made to recognize and acknowledge the reality of heterogeneity—and there have been a multitude of such attempts, especially in the last two centuries (rang-ing from Hegel's to Bourdieu's)—they have almost exclusively been aimed at showing the underlying unity of what merely appeared to be diverse. Yet no consensus has been reached on what that unity might be.

An examination of "interpretive communities," whether of the American pragmatist persuasion or the more recent post-Heideggerian stylizations, shows us that such communities pay no serious attention to one another. For communities of discourse, mutual engagement is fundamentally an internal matter (and a highly fractured one at that). Thus, for example, while there is an ongoing effort to disprove Freudian theory, most of those who use it don't care; they continue to analyze patients, movies, etcetera in Freudian terms. And the overwhelming majority of literate, or semiliterate, knowledge pro-ducers, who have never read a word of Freud, don't care either. As there are no sanctions except mutual contempt and the nasty book or grant review, this situation is unlikely to change. Different interpretive "federations," or simply clusters, coalesce around different questions, different methods, different standards of evidence, different types of argumentation, different career patterns, different sources of symbolic capital, differential placements within the cultural, economic, political, and social fields. Then such clusters themselves produce other subclusters, and discursive battles ensue.

This state of things is partially the result of the fact that within the human sciences no stable mechanism has been invented to centralize policing, to enforce "order and progress," to cite the old positivist motto. To make a long argument short, in the natural sciences the academies and granting agencies function as gatekeepers; without money and facilities there is no natural science. In the human sciences, no such mechanisms exist, or none, at least, approaching the same effectiveness. As salaries continue to be paid, discourses continue to augment. Only in authoritarian systems has a degree of consensus been reached and sustained. This claim extends from the hard authoritarianisms like communism or National Socialism to softer ones in which elites rule by habitus and class affiliation and thereby control boundaries through appointments and commissions alone.

So what is one to make of this dissonance? One way out is to adopt a meta-position that begins with a principled affirmation of the inevitable plurality of positions. An inevitable plurality of *logoi* and perhaps of *ethē* as well. Philosophers in the American pragmatist tradition made a number of attempts to think this state of pluralism through as a positive condition of thought and value. From John Dewey through Richard McKeon, they have provided significant reflections on maximizing the utility and public good attained through an acknowledgment and affirmation of pluralism. Their positions, however, have tended to constitute themselves as schools and have encountered eventually the same types of divisions and disputes as other schools of philosophy.

Equipment

Why, when it comes to thinking, is there this vexation? This irritation, this distress, this tossing about? Although logos, reasoned discourse, must be a part of the solution, as what we are doing is thinking, it seems also to constitute an essential dimension of the problem. This insight might lead one to conclude that logos is expendable. Nothing could be farther from my project. Rather, it seems to me that the starting point of inquiry and reflection, the anthropological problem, lies in the apparently unavoidable fact that anthropos is that being who suffers from too many logoi.

To say that relating logos to ethos is problematic is to rephrase what has just been said. Attempts to establish a relationship between these two terms have produced different affects. Among these affects is pathos.[5] Remembering that pathos is both a medical and a theatrical term, its presence can be taken up as both diagnostic and representative. Its presence is diagnostic in the sense that something seems wrong: a form of care is called for. The presence of pathos is representative in that all staging of answers themselves eventually pose the problem of how something can be represented. It follows that an attention to form is inescapable.

When Foucault undertook his famous detour into ethics during the 1980s, the topics of care and form became central. He turned to the genealogy of a type of relation between thinking and acting to which he had not given prominence, a relation that was pragmatic but not immediately political. He entitled his course at the Collège de France during the academic year 1981–82 "L'herméneutique du sujet." The course was devoted to exploring the techniques, practices, and reflections related to "care of the self" in the late antique world. The guiding hypothesis of Foucault's rich and far-reaching lectures was that for almost two millennia the imperative to take "care of the self" had been linked to, and in fact primed, the imperative to "know thyself." Knowledge was not an end in itself; it was an essential element of a life well led. Its function was to contribute to such a life.

In the early 1980s, Foucault devoted himself to archaeological explorations of the sundering of the imperative to "know thyself" from that of its lost partner, the "care of the self." The genealogical dimension of Foucault's work explored the possibility of recreating this alliance as a problem of actuality—not, of course, to return to the older solutions but to find among those solutions a way of formulating a contemporary problem with more clarity. Frédéric Gros, the scholarly editor of the 1981–82 course, in his excellent "Afterword," succinctly sums up the core of Foucault's concerns as follows:

(1) Can one have access to truth without putting into question the very being of the subject who achieves that access? Can one have access to truth without paying the price of a sacrifice, of an asceticism, of a transformation, of a purification, that touches the very being of the subject?

(2) Can self knowledge, understood as part of knowledge in general, take account of the care of the self?[6]

There existed in late antiquity a corpus of arts and techniques considered by all to be essential to the care of the self. Much of Foucault's inquiry in the 1981–82 lectures focused on this corpus, these practices, these exercises, constituent of, and essential to, self-formation and care. His preliminary working hypothesis was that in the Western philosophical tradition there had been three major forms of reflexivity. By reflexivity Foucault means exercises of thought in which the act of thinking is itself made an object of thought. The three forms were memory, meditation, and method. In this instance, as elsewhere, Foucault is using terms such as "memory" or "method" as topics to begin an inquiry. He starts by taking one of these terms—for example, "memory"—focusing on one exemplary use of it in the writings of Plato. He then analyzes the constituent elements of the exemplary case. The recombination of these elements, as well as the addition of new ones, provides the material means to articulate a space of variation and development. This space is not the historian's space. Rather, it is a logical space, composed of historically defined and situated elements, close to that of Max Weber's "ideal types" (as we shall see later).

It was with the emergence of program of method as certitude that the concerns with the ethical conduct of a life were sundered from the search for truth. Method was conceived as operating as a form of objectivity and autonomy. Method was *amoral* in the sense that the subject of knowledge no longer needed to be in a privileged ethical state to receive the truth.[7] And the reception of objective truth had no necessary consequences for the ethical state of the subject who received it. The search for a method is a search for a "form of reflexivity that seeks a certitude that can serve as a criterion for all possible truths, and which, from that fixed point, can lead truth to a systematic organization of an objective knowledge."[8] Both of these forms, memory and method, are well known, even if their histories and fates have been complex. Neither memory nor method, however, is at the heart of Foucault's analysis. Rather, they are topics that enable him to better define the space of "meditation."

What is meditation? In the late antique world, meditation differed profoundly in its goals, practices, and forms, from meditation today. Today "meditation" carries the connotation of either an attention to inward states or of attempts to empty the mind. The gulf that separates the older uses of this term from the current ones stands out in the definition Foucault provides. "The test of one self as a thinking subject, who acts and thinks accordingly, who has as his goal a certain transformation of the subject such that there is a self-constitution as an ethical subject of truth."[9] Meditation, then, was an exercise, an exercise of thought directed to thinking, an exercise whose goal was to connect thought to ethos.

One of the characteristic ways of describing the care of the self was as a set of exercises that prepared one for a lifelong battle against external events. Sometimes this preparation, and its associated exercises, was described as an athletic contest, sometimes as a battle. In either case, one needed a supply of proverbial weapons in order to endure and to triumph in the conflict. Foucault captures this dimension in one of those invigorating turns of phrase at which he was so gifted: "The Stoic athlete . . . had to be prepared for a battle, a battle in which his adversary was anything that might come at him from the outside world: the event. The antique athlete is an athlete of the event. As for the Christian, he is an athlete who confronts himself."[10] One needed a training in vigilance and agonism, because these tests were challenges at which one sought to excel, not merely to triumph or survive.

Further inquiry would reveal that historically these types would have been broken down into elements and these elements recombined in various manners. They would have been rethought and put to different uses in different contexts. Thus, for example, centuries later, work on the self, even the interior self, would come to be understood as coping with the inner significance of events. Such work, of course, would have its own distinctive practice of memory and method.

The care of the self, then, was not just a state of consciousness; it was an activity. Furthermore it was not an activity appropriate just for this or that occasion; rather, it was an essential dimension of a whole way of life. It was a constitutive element of a form of life. Thus, in one sense it was part of a broader pedagogy, in the ancient sense of *paideia*, or in the more modern sense of *Bildung*. However, the care of the self was more than that; it was more than a stage (or set of stages) one passed through. The care of the self was also a form of critique, a critique of the self that entailed perpetual self-examination, an unlearning of bad habits as well as the forming of good ones. In sum, meditation, *meletē*, was an exercise in the practical appropriation of thinking about and toward the self. It was an appropriation aimed at literally forming the subject. It was not aimed at merely enriching his knowledge, building his reputation, or polishing his style for its own sake. The care of the self was an essential aspect of how a moral existence had to be lived. Although this preparation and this exercise focused on the care of the self, it was far from being a solitary affair. In fact, the practice of the care of the self passed through an elaborate network of relationships with others. The care of the self was highly social, and it was oriented from the self outward to others, to things, to events, and then back to the self.

How was this work, how were these exercises, to be accomplished? In the late antique world there existed a whole range of "equipment" to aid those engaged in these exercises. The key "equipment" that was required to take care of the self, to aid it in its confrontations with the proverbial slings and arrows of the external world, or more generally to accomplish the complex task of facing the future, was an arsenal, if you will, of logoi.[11] This inventory of logoi formed a kind of tool chest. The Greek word for this toolkit is *paraskeuē*, or "equipment." As the name suggests, this equipment was designed to achieve a practical end. These "true discourses," these "logoi," were neither abstractions nor, as we say today, "merely discursive." They had their own materiality, their own concreteness, their own consistency.

What was at stake in the use of this equipment was not primarily a quest for truth about the world or the self. Rather, it was a question of assimilating these true discourses, in an almost physiological sense, as aids in confronting and coping with external events and internal passions. The challenge was not just to learn these maxims, often banal in themselves, but to make them an embodied dimension of one's existence. To have them ready at hand when needed: "to make of a taught, learned, repeated, and assimilated logos the spontaneous form of the acting subject."[12] True discourses were equipment to the extent that they had been assimilated thoroughly, made to function as rational principles of action. Learning these maxims was not hard; accomplishing the goal of making these logoi a principle of action was a lifelong process.

Throughout late antiquity, Epicureans, Stoics, and Cynics ardently debated the best use of this arsenal of logoi within the problematic of the care of the self. But all the schools of thought agreed on two things: (1) that care of the self and knowledge of the self went together, with the former priming the latter; and (2) that the deployment of true discourses was absolutely not a question of deciphering the hidden meaning of our thoughts and desires.[13] Thought was inseparable from the world, from the self, from others, from events. Thought was a practice. In sum, "*paraskeuē* . . . is the medium through which logos is transformed into ethos."[14] The challenge of bringing logos and ethos into the right relationship was, and is, the challenge confronting anthropos.

Modern Equipment

In *French Modern: Norms and Forms of the Social Environment*, I traced some of the dimensions of how modern urban planning had gradually developed over the course of the nineteenth century. Urban planning had started with the rational reform of physical space but had gradually included more and more elements in its purview. By the time such planning had become a socialist project during the 1930s, it was proud of having expanded its scope from city planning, *un plan de ville*, to planning that included all those elements (spatial, social, psychological, architectural, hygienic, etc.) that contributed to shaping an individual life, *un plan de vie*. The goal of planning was social and individual health, a well-policed order. By 1942, the French "Plan d'Equipement National" defined *équipement* as everything that was not a "free gift of the soil, subsoil or climate. It is the work of each day and the country as a whole." One could say "equipment" had become the subject matter of method. In a parallel fashion, one could say that the subject had equally become an object of method.

Thus, viewed from our current perspective, we could say that a tool chest of logoi had been gradually assembled, and partially put into practice by the state. Further, new social technologies had been invented to oblige individuals to have these rational aids ready at hand on all occasions; or, failing that, at least to have social specialists nearby who could bring the corrective benefits of these technologies to bear with the shortest possible delay. The political rationality consisted in recuperating and subsuming, through method, the traditional functions of meditation.

The task of this book is neither to rehearse the archaeology of these changes nor to evaluate them. Rather, what I am attempting to do is to reflect on how it might be possible to transfigure elements of the equipment of modern method into a form of modern meditation, and to bring the benefits and effects of that transformation to bear on inquiry. The challenge is threefold: (1) to provide a toolkit of concepts for conducting inquiries into the contemporary world in its actuality; (2) to conduct those inquiries in a manner that makes the relations, connections, and disjunctions between logos and ethos apparent and available to oneself and to others, that is to say, to make those relations part of the inquiry itself as well as part of a life; (3) to take into account the pathos encountered and engendered by such an undertaking, and to find a place for it within the form under construction.

Form

The German philosopher Peter Sloterdijk, in a lecture entitled "Règles pour le parc humain" ("Rules for the Human Theme Park"), provides a diagnosis of the current prospects for the human condition. The 1999 lecture occasioned a violent polemic, first in Germany and subsequently in France. The polemic turned on a future policy of genetic selection that Sloterdijk purportedly advocated. Like most polemic, the debate was only tangentially concerned with the truth of what had actually been claimed. As Sloterdijk points out in a "postface" to the French edition of his lecture, the distinction between description and prescription had escaped his critics. In fact, one would never have guessed from reading the accounts of the battle that the bulk of Sloterdijk's lecture was devoted to a kind of genealogy of humanism in Western culture.[15]

The lecture has three parts: an overview of humanism understood as rooted in written exchanges between friends; a reflection on Martin Heidegger's "Letter on Humanism"; and a diagnosis of a possible coming age in which genetic "selection" would replace other institutions as a basic form of domestication. I find the last two sections of minor interest and distance myself from Sloterdijk's formulation of the problem as well as his answers. In the first part of his lecture, Sloterdijk presents the thesis that for well over two millennia Western philosophy has taken a specific form. That form was an exchange of letters among friends. As Jean Paul wrote, books were essentially letters written to friends. This form established itself when philosophy as an oral exchange between citizens of the same polis came to an end, that is, when Greek thought was transferred to Rome. With the change of site the privileged medium became writing. These facts are banal and can be found in any standard textbook.

Where Sloterdijk takes his tale into more fertile territory is with his claim that "Philosophy is the ability to make friends through the medium of a written text."[16] This friend-making was no longer the one Aristotle valued, the long-term relationships, tested and developed in the face-to-face space of the polis. The newer form bridged geographical separations and was conducted over the course of generations. Philosophy, friendship, and writing were elements of a form that required and made possible a series of deferred exchanges among the living and the dead. A central characteristic of this form of epistle devoted to reflection and addressed to another is that while the original letter might well be sent to a specific person, it could be read across time and space by many others. Writers lived with the knowledge that their work was capable of producing an "indeterminate quantity of possibilities to establish a link of friendship with readers who are unknown and often not even born."[17] One might say that the horizon of philosophy was without temporal or spatial borders. Rather, it was social status that separated and joined individuals; the price of entry into the game of philosophy was leisure and literacy, although obviously not all those who had the means to do so chose to play the game. The practice equally required a curiosity and a receptivity. Furthermore, it did not actually require the participants to "communicate" with each other in any direct sense.[18]

The second aspect of Sloterdijk's claim is that this form had served a specific function. The function was to "domesticate" [*apprivoiser*] humanity as instantiated in the self. Western humanism has always opposed itself to—and thereby connected with the possibility of itself transforming into—one or another form of barbarism. The role, the task, the challenge of humanism was to overcome, or at least to tame, the barbaric. Among the elite philosophers—the men of leisure and learning—the challenge was to live a cultured life, that is, a moral life, among a barbaric populace policed and entertained by an ultimately brutal civilization. The challenge was to overcome that barbarity through work on the self and work on, and with, those others that counted for oneself.

What later humanists understood as *humanitas* obliged them to adopt a distance from the entertainment of the masses "in the theaters of cruelty. Thus even when the humanist wandered among the howling crowd, it was only to underscore that he too was a human being who could be infected by bestiality."[19] The path out of and away from that bestialization was through what Sloterdijk calls a choice and correct use of "media." By "media" he means the form given to the practices of work on the self and others. Those practices centrally included writing. Foucault has given us a rich account of some of those practices in volumes 2 and 3 of *The History of Sexuality*.

Leaping forward a millennium and a half, Sloterdijk lands in the project of nation building as the challenge of culture and civilization. He asserts—again a standard textbook claim—that with the generalization of literacy between 1789 and 1945, the function of literary form changed. In a wonderful turn of phrase, Sloterdijk writes, "Populations were organized as obligatory fellowships [*des amicales obligatoires*], thoroughly literate and civilized by an obligatory canon of readings within a national space."[20] One epic battle of the ancients and the moderns was over: one could now have both the classical canon and a new one at the same time. Nationalism could incorporate them both into one body of learning, one technology of civilization.

The decline of this second form of literacy and cultivation—this "obligatory fellowship"—has many causes, all stemming from the end of the absolute dominance of the nation-state as the form of sovereignty. Central among those causes, Sloterdijk underscores, is the rise of mass society and its integral connection to new media: with radio (1918), television (after 1945), and the Internet (1989), a new world of post-literary, post-epistolary, and post-humanistic culture has inexorably spread its web. Almost as if hearing himself read what he had just written—perhaps one can become a philosophic friend to oneself—Sloterdijk retreats just a little. The great modern societies, he observes, can now only "marginally" produce their political and cultural syntheses through traditional forms of media.[21] Neither literature nor philosophy have disappeared; it is only that they have ceased to play the central civilizing role for the nation-state. These practices are still present, but they are now located at the margins of the civilizing process.

The second part of Sloterdijk's lecture is devoted to Heidegger and his "Letter on Humanism." The one aspect relevant to this discussion is Sloterdijk's rejection of Heidegger's absolute line of demarcation between humans (Dasein) and animals as the basis for any new humanism. Humans are the only beings who live in the clearing of language. This dwelling is their essence. It radically differentiates humans from other living beings. The cultural world is not the environment. Sloterdijk's diagnosis is that in the current clearing of being it will no longer be possible to skirt the question of the biological constituents of human life—and the role they should play in the civilizing or enculturation process. I agree that a form of this question is certainly confronting us. However, I see no compelling reason whatsoever to reintroduce all the complications and obscurity that Heidegger's concept of the "clearing" brings with it. Furthermore, Heidegger's "Letter" can hardly serve as an exemplar of philosophic exchange. Neither its tone, nor its mission, nor its contents lead in a desirable direction. Rather, Heidegger, and unfortunately Sloterdijk as well, write more in the tradition of prophecy.

Discontents and Consolations

The rosy blush of . . . the Enlightenment, seems also to be irretrievably fading, and the idea of duty in one's calling prowls about in our lives like the ghost of dead religious beliefs.

Max Weber[22]

During the course of his essay Sloterdijk asks what seem to me to be two rather different questions, each addressed to a particular kind of problem. At one point Sloterdijk asks whether there is still a "dignity of the human being which merits expression in philosophic reflection."[23] However, earlier in his text, Sloterdijk had asked a rather different question, a question that does not, it seems to me, presuppose the form of possible answers: What form could be available through which humans could become humans by overcoming their brutal and bestial impulses? That question, Sloterdijk observes, "implies nothing less than an *anthropo-dike*—that is to say, a determination of the human being as concerns its biological incompletion and its moral ambivalence."[24] The following chapters approach this topic from different angles. These orthogonal approaches do not seek to answer the question but to provide elements that could help to pose it better.

Discontents

In 1930, Sigmund Freud, already in a sombre, pessimistic mood about the state of the world (a mood reinforced shortly thereafter by the victories of the National Socialists), published *Civilization and Its Discontents*. Perhaps defiantly, Freud conspicuously continued the scientifically detached stance he had fashioned in *The Future of an Illusion*. This stance, with its resigned distance, and its self-control, was both the price to be paid and the constraint required, or so it seemed to Freud, to pursue successfully the project of de-mystifying humankind's deepest illusions. By means of this ascetic exercise, Freud believed he could achieve, or had already achieved essential insights that others, mired in illusion, lacked. That lack—Freud was lucid about this point—provided its own benefits in the world, benefits that those pursuing science would have to forgo as the price of insight. Basically, for Freud, what had to be abandoned was hope, or at least naïve hope.

It would seem to follow that abandoning childlike hope was a necessary, if not definitive, step toward maturity, or perhaps wisdom. But is there such a thing as scientific maturity or wisdom? Much turns on the term *Wissenschaft*, science, and what it offered, and to whom. I take Freud's claims and the position he claimed them from as a starting point for exploration, with the hope that such an effort might help us renew *Wissenschaft* through a commitment to making it a central component of a life.

One of Freud's central claims was that mankind for most of its history had unknowingly projected its ideals onto its gods. Recent advances in civilization, however, had complicated this age-old process; not only were some of these projective processes now understood (thanks to the scientific advances Freud himself was spearheading), but additionally, and this was more complicated yet, mankind was close to making its ideals into realities. "Man has, as it were, become a kind of prosthetic God." This double turning of increased self-awareness and increased power constituted the diacritic of the present. What Freud held to be certain was, first, that the process would continue indefinitely into the future; and second, that "present-day man does not feel happy in his Godlike character."[25] And humans, according to Freud, believe that they desire to be happy. Consequently, discontent was another diacritic of the human plight, especially as science advanced and its achievements yielded instrumental capacities. Freud's prognosis, in 1930, was gloomy. While scientific and technical advances were unquestionably accumulating, the contemporary mix of scientifically achieved self-understanding (of the self and of civilization) and technical advance was not yet coordinated. Mankind was pursuing its illusions with more power than ever before. Freud's effort was to question the project of coordination or at least to temper the expectations it engendered. Of course, Freud himself was deeply committed to a scientific project of his own.

In 1916, a younger Sigmund Freud had written a small article for a Hungarian journal entitled "A Difficulty in the Path of Psycho-Analysis." The piece, which appeared early in 1917, was intended for an "educated but uninstructed audience" (an interesting distinction when you think about it). Freud remained content with the article's basic points and repeated them, albeit phrased a little differently, in his subsequent *Introductory Lectures in Psychoanalysis*.[26] The difficulty alluded to in the title of Freud's essay was not humanity itself but rather its pride. (The question of who exactly this humanity or mankind is, it is worth remarking, is not explored in the essay.) Freud's core argument is that throughout history scientific advance had run counter to humanity's megalomania, its self-importance. Consequently, it was consistent to assume that any truly significant scientific advance concerning man's relation to the cosmos, to nature, to other humans, or to himself would be resisted, for longer or shorter periods of time.[27] Freud's core position is that as science discovered and demonstrated what was true, mankind ultimately had no rational alternative but to adapt its own self-understanding to scientific discoveries. In these articles, as elsewhere, Freud presents himself as a scientist, even a great scientist; this self-presentation constitutes an audacious challenge to his readers to accept his theories and no doubt offers some comfort to himself. After all, the article was written to explain why his theories were not being generally accepted.

Furthermore, in his defiant faith in the inevitable triumph of science against the blind forces of irrational resistance to its discoveries, Freud can

be seen as not merely a scientist but an *Aufklärer*, a man of the Enlightenment. The distinction rests on the observation that there is nothing within
the disciplinary confines of this or that science to direct the historical fate
of its discoveries. An *Aufklärer* is someone who undertakes to pursue increased understanding of a rational sort wherever it leads, believing that it
will lead somewhere beneficial. Enlightenment affect (belief, hope, desire)
is a surplus, a supplement, to scientific achievement. An *Aufklärer* follows
Kant's dictum, *sapere audere!*—"Dare to know!"[28] As Kant argued, enlightenment is simultaneously a scientific, moral, and political undertaking.
Such a project constitutes a commitment to a kind of truth and to a way
of life linked to an understanding of the good. Enlightenment, one might
say, is a culture, an ethos, or a form of life. It is a form of life that can never
be complete. It is a form of life that is both arrogant and humble. It is arrogant insofar as it acts for humanity with a confidence that it is right; it is
humble in that enlightenment is an infinite project whose achievement lies
in the future.

Consequently, an ethos of enlightenment is a way of life that requires
a certain understanding of maturity, that is to say, a view of the past, the
future, and the present that links them together in a hopeful manner but
the proof of which can only lie in the future of humanity, not in any individual life. The question is whether there is a corresponding ethos within a
scientific attitude. I will recurrently raise the issue of maturity and its relation to science, enlightenment, and history. The reason for this repetition is
that there are different and contrastive understandings of each of the terms.
Those differences depend in part on an evaluation of the history of science
and enlightenment—and of the present moment.

Freud proposes "to describe how the universal narcissism of men, their
self-love, has up to the present suffered three severe blows from the researches of science."

1. The cosmological blow. Man believed that his abode, the earth, was
   the stationary center of the universe. This perception fit well with
   man's "inclination to regard himself as lord of the world." The first
   blow to mankind's lordly status was dealt when it learned that the
   earth was not the center of the universe but only a tiny fragment of
   a cosmic system of scarcely imaginable vastness. The destruction of

this narcissistic illusion came to general acceptance in the sixteenth century with Copernicus, although Freud is at pains to underscore that the discovery had been made millennia before.

2. The biological blow. In the course of the development of civilization man acquired a dominating position over his fellow-creatures in the animal kingdom. Not content with this supremacy, however, he began to posit a gulf between his nature and theirs. He denied the possession of reason to them and to himself he attributed an immortal soul, and made claims to a divine descent that permitted him to break the bonds of community between himself and the animal kingdom. Darwin put an end to this presumption. "Man is not a being different from animals or superior to them; he himself is of animal descent, being more closely related to some species and more distantly to others." Although this point has been hard for civilized adults to accept, Freud insists that children and primitives readily assume a closeness with animals.

3. The psychological blow. This, in Freud's self-serving opinion, is probably the most wounding. Man has already been humbled externally but now must accept that he is not sovereign even within his own mind. Philosophers had previously understood this point, but its scientific demonstration has nevertheless been fiercely resisted. Man, it seems, must also accept that he is thinking about sex all the time, and that only Sigmund Freud has explained why.

Regardless of how one evaluates Freud's overall thesis, it is important to note that he does not explain under what historical conditions scientific truth becomes socially acceptable, or even address this question in any way. Greek scientists knew the earth traveled around the sun, children feel a kinship with animals, and philosophers knew that we know not what we think. But all these truths were resisted by the culture at large. Yet somehow, eventually, even grown-up Europeans saw, and would see, the light of day. In this faith, despite all his pessimism about civilization and its discontents, Freud remains an enlightenment thinker. Not only does he dare to know— fulfilling the highest commandment—but he assumes that ultimately the truth will, as it were, come to light. That light, sooner or later, will shine forth, and humanity will awaken. Was Freud's faith his ultimate defense mechanism, or a sign of his maturity, a maturity running ahead of the rest of mankind and presaging where it is heading?

Science as a Vocation: Truth Versus Meaning

In 1917, perhaps on November 17, the very day of the Bolshevik seizure of power in Russia, Max Weber delivered a lecture, "Wissenschaft als Beruf" ("Science as a Vocation"), to a crowded hall of German university students in Munich.[29] It stands as one of the great—unsurpassed, in my view— twentieth-century statements of the ethics and ethos of science and scientists. It may well be considered one of the first twentieth-century statements, especially if one agrees with my old German humanist professors at the University of Chicago, who felt that Western civilization had come to an end by 1917. The lecture fits within the general framework Weber had elsewhere set for himself of characterizing the "life orders" (*lebensordnungen*) under modern capitalism. Although Weber does not phrase it this way, the central theme of the lecture might well be epitomized as, What is maturity, within modernity, for those who dedicate their life to seeking knowledge and understanding? In the triad of science, enlightenment, and history, Weber privileges history and science. He presents a challenging diagnosis of the historical moment and the ethical demands it poses for those who desire to remain loyal to science. Loyal, that is, without illusions. Weber chillingly refers to the enlightenment as "the laughing heir" of capitalism, an heir that, by 1917, had long lost its "rosy blush."[30] For Weber, we lived enmeshed in processes of modernity rather than enlightenment.

Weber divided his lecture into three parts: (1) the material conditions of science, (2) the inner ethic of science, and (3) the value, or cultural significance, of science in modernity. Although this set of distinctions is totally out of fashion today, I believe it remains a powerful mode of orientation for those who study science and practice *Wissenschaft*.

Material Conditions

Weber cast his discussion of the material conditions of science as a comparison between the work conditions and career trajectories of graduate students in Germany and the United States. German students, after a lengthy apprenticeship and the publication of a book, received permission to begin offering lectures, for which they were compensated only by the fees of those who attended. While providing limited monetary resources, this system left the student a good deal of freedom of thought and time to conduct research. In the United States, by contrast, an academic career began with a regular faculty position. Hence the young person joined a bureaucratic system and was assured of being paid, though often, Weber observes dryly, the equivalent of the wages of a semiskilled laborer. (Only football coaches were well paid in American universities, Weber observed.) In return for this position and the modest level of financial security that came with it, the young scientist was required to do a great deal of teaching, although ultimately his career would be judged on his research. Whatever else it might be, *Wissenschaft*, for Weber, required labor and institutional resources.

With a certain regret, Weber observed that the old humanist university in Germany was on its last legs.

> In very important respects German university life is being Americanized, as is German life in general. . . . The large institutes of medicine or natural science are "state capitalist" enterprises, which cannot be managed without considerable funds. [As in all such enterprises, there is a separation] of the worker from his means of production. The worker, that is, the assistant, is dependent upon the implements the state puts at his disposal; hence he is just as dependent . . . as is the employee in a factory upon the management. . . . As with all capitalist, and at the same time bureaucratized, enterprises, there are indubitable advantages in all this.[31]

And disadvantages. Not only was science operating under capitalist and bureaucratic constraints, it further labored, like the Vatican, under conditions of consensus formation that rarely rewarded exceptional people. Weber paints a stinging, and remarkably contemporary, portrait of the role played by chance, arbitrariness, and consensus formation in academic life. "It would be unfair to hold the personal inferiority of faculty members or educational ministries responsible for the fact that so many mediocrities play an eminent role at the universities. The predominance of mediocrity is rather due to the laws of human cooperation."[32] Consequently, a young person contemplating a scientific or scholarly future must ask himself, "Do you in all conscience believe that you can stand seeing mediocrity after mediocrity, year after year, climb beyond you, without becoming embittered and without coming to grief?"[33] Enthusiastic young people always answer that their "calling" for science will see them through, Weber remarked, but few actually make it, without succumbing to ressentiment or resignation.

Finally, not all were allowed to play the game of science. Weber does not mention gender, even though his wife was an ardent socialist-feminist, but he does add that if the would-be scientist was "a Jew, of course one says *lasciate ogni speranza* [abandon all hope]."[34] This equation of the gates of *Wissenschaft* with the gates of hell is, upon reflection, a rather bizarre one. It should serve as a lesson to those who pine for the good old days when science was pure. By this I do not mean that the recent couplings of science and industry are unproblematic, only that historically their separation contributed to a certain castelike recruitment within Germany and beyond.

Inner Ethic

Weber opens the section on the "inward calling for science" by continuing to specify the conditions under which science operates. The essential feature of contemporary science is that it has entered an irreversible "phase of specialization previously unknown, and that this will forever remain the case."[35] Science is not wisdom, science is specialized knowledge. A number of important consequences follow from this situation. First, "scientific work is chained to the course of progress."[36] All scientists know that, by definition and in part due to their own efforts, their work is destined to be outdated. Every scientific achievement opens new questions. One might say that a successful scientist can only hope that his or her work will be productively and fruitfully outmoded rather than merely forgotten. Second, the knowledge worker must live with the realization that not only are specialized advances the only ones possible but that even small accretions require massive dedication to produce. Dedication or enthusiasm alone, however, are not sufficient to produce good science, nor does hard work guarantee success. "Ideas occur to us when they please, not when it pleases us."[37] The calling for science thus must include a sense of passionate commitment, combined with methodical labor and a kind of almost mystical passivity or openness. The scientific self must be resolutely willful and persistent, yet permeable. Androgynous, if you will.

Here Weber opens a parenthesis that is one of the most celebrated in his entire work. What exactly, he asks, does scientific progress provide to the individual, to society, and to civilization? His answer is a stark one: science alone does not produce either enlightenment or meaning; in fact, under conditions of modernity, science stands in a fraught, perhaps mortal, tension with both enlightenment and meaning. For Weber, scientific work forms part of a larger "process of intellectualization" that has been developing for thousands of years. What does this mean?

> Does it mean that we, today, for instance, . . . have a greater knowledge of the conditions of life under which we exist than has an American Indian or a Hottentot? Hardly. Unless he is a physicist, one who rides on the streetcar has no idea how the car happened to get into motion. And he does not need to know. [He can depend on others.] The savage knows incomparably more about his tools. . . . The savage knows what he does in order to get his daily food and which institutions serve him in this pursuit. The increasing intellectualization and rationalization do not, therefore, indicate an increased and general knowledge of the conditions under which one lives. It means something else, namely, the knowledge or belief that if one but wished, one could learn it at any time. Hence, it means that principally there are no mysterious incalculable forces that come into play, but rather that one can, in principle, master all things by calculation. This means that the world is disenchanted. One need no longer have recourse to magical means in order to master or implore the spirits. . . . Technical means and calculations perform the service.[38]

"Now, this process of disenchantment or de-magification, which has been unfolding in Western culture, [does it] have any meanings that go beyond the purely practical and technical?"[39] Strictly speaking, within the constraints of the issue of the "inward calling for science," there can be no answer to this question, because it can not be addressed scientifically. If we recall that when Weber refers to *Wissenschaft* he means all forms of disciplined knowledge, we are unlikely to be let off the hook by bringing Shakespeare to the physicians nor ethics committees to the molecular biologists. For that move risks instrumentalizing the cultural sciences (*Geisteswissenschaften*) rather than humanizing the life sciences.

The Value of Science

"To raise this question is to ask for the vocation of science within the total life of humanity."[40] The mission of science is quite specific: to invent concepts and conduct rational experiments. These concepts, however, no longer provide a window onto eternal verities, and the experiments no longer reveal absolute truth. Furthermore, they tell us nothing about the meaning of the cosmos, nature, or the psyche. Weber heaps scorn upon those who think otherwise. "And today?" he scoffs. "Who—aside from certain big children who are indeed found in the natural sciences—still believes that the findings of astronomy, biology, physics, or chemistry could teach us anything about the meaning of the world?"[41] Or, "After Nietzsche's devastating criticism of the 'last men' who 'invented happiness,' I may leave aside altogether the naïve optimism in which science—that is, the technique of mastering life which rests upon science—has been celebrated as the way to happiness. Who believes in this?—aside from a few big children in university chairs or editorial offices."[42] Or, "Natural science gives us an answer to the question of what we must do if we wish to master life technically. It leaves quite aside, or assumes for its purposes, whether we should and do wish to master life technically and whether it ultimately makes sense to do so."[43] Weber shares with Freud the view that science and its associated growth of instrumental capacities was not the path to happiness. He differs from Freud in refusing to believe that scientific truths yielded meaning. For Weber, science alone could not yield meaning; the only possible path toward that goal was experience yielding phronesis. Weber deeply desires to follow this path but despairs of making any progress in doing so.

For Weber, science contributes methods of thinking, the tool and the training for disciplined thought. It contributes to gaining clarity; that is all. Hence, for Weber, science contributes to an ethics; a critical ethos of "self-clarification and a sense of responsibility." This sense of responsibility turns on a specific conception of truth. Such an ethics is a form of critique, in the Kantian sense of establishing where the limits of thought lie. It is also critical in the sense that it displays a suitable scorn for those who cannot accept the limits of what *Wissenschaft* provides. That science "does not give an answer to questions [of meaning] is indubitable." On that claim Weber brooked no gainsaying. However, that insight constituted not the end but only the beginning of the problem of science, ethics, and modernity. "The only question that remains," Weber continued, "is the sense in which science gives 'no' answer, and whether or not science might yet be of some use to one who puts the question correctly."[44] In the conclusion, we will return to Weber's far-reaching, still unanswered, and entirely contemporary query.

However, today it seems clear that Weber's views of history and of science (*Wissenschaft*) require modification. Specifically, they are too monotone and too substantialist. At times Weber remains a neo-Kantian, seemingly forcing science into a priori categories. At other times, he seems almost to hold a view of "rationalization" as the master term of Western history (although in other places he resists this hypostatization). Both tendencies go against the grain of other aspects of Weber's thought, in which categories such as science seem more like ideal types, hence become an analytic focus relative to particular value orientations and are historical and contingent. Wherever one comes down in these debates, Weber's question and concern about the status and challenge of the life orders within modernity, it seems to me, remains a compelling one, even if his answers seem dated.

1917–1989: Enlightenment Betrayed

The twentieth century, amply endowed with megalomaniac projects, has been the scene of further wounds to mankind's naïveté and its narcissism. The ever-reasonable, prudent, and cautiously hopeful Jürgen Habermas observes, "historical scepticism about reason belongs more to the nineteenth century, and it was not until the twentieth century that intellectuals engaged in the gravest betrayals."[45] Although Habermas is presumably referring to intellectuals such as Martin Heidegger (and his obscene allegiance to the Nazis) and Georg Lukács (and his horrific indentureship to Stalin), his remark applies to natural scientists as well. The twentieth century witnessed the establishment of a potent and malign connection between knowledge and the military (and of forces of destruction more generally), from the horrific effects of poison gas (and other gifts of the chemical industries), through the atomic bomb (and other gifts of physics and engineering), through the Nazi nightmare of racial purification (and other gifts of anthropology and the biosciences), to the indigestible fact that close to three-quarters of the spending on scientific research during the Cold War was devoted to military ends. The industries and sciences of thanatos have had a glorious century. We should never forget that what is nostalgically seen today as the golden age of science—the one before capitalism supposedly despoiled the life sciences—was the age of the Cold War. Today it seems implausible to maintain any longer that accumulating knowledge per se automatically leads to beneficial results, or, given its fragmentation, furthers our general self-understanding. Nor can we—and this is where Weber helps us avoid the fatuous denunciatory cant so widespread today—equivocally maintain that the opposite is the case, that is, that science is malign and darkens our self-understanding.

It is striking that in *The Human Condition* (1958), Hannah Arendt chose physics as her exemplary science. So did C. P. Snow in *The Two Cultures*, published the same year. However, four years later, in the book's second edition, Snow replaced physics with molecular biology. He was prescient. The immense achievements of molecular biology and biochemistry during the 1960s and 1970s—the discovery of the fundamental principles and mechanisms of the genetic code and its operation—will surely stand as a monumental threshold in the history of science. However, with the development of recombinant DNA technology and the emergence of a new type of industry—the biotechnology industry—another blow was dealt to those who wanted to believe that the production of knowledge about "life" must remain pure of worldly taint. Over the last two decades it has been shown that there can be no life science without substantial amounts of money. During the Cold War this money came from nation-states. Although the State still contributes substantially to the life sciences, an even greater flow of funds issues from the huge multinational pharmaceutical companies and from the fleet-footed and highly mobile purveyors of venture capital. Please note that I am not claiming that this situation is intrinsically either horrific or wonderful: I have no regrets for the cessation of the Cold War, or for much of what nationalistic science produced in the twentieth century. I have no doubt that the goals and means of capitalist enterprise and character will inflect, perhaps radically, what used to be known as the scientific ethic. My goal is simply to note a watershed change and to urge us to reflect on it.

Although hype and cant have dominated the coverage of the emergence of genome mapping, what we have learned from the first decade or so of this enterprise are neither the secrets of the Holy Grail of life nor the meaning of its Code of Codes. Nor has it been demonstrated that genetics inevitably brings with it a new eugenics. Rather, we have learned that all living beings, at the level of the genetic code, are materially the same and that the very techniques that were developed to make this profound discovery enable, even oblige, further intervention into that materiality. François Jacob, the French Nobel Prize winner, frames these two points in simple, elegant prose: First, "All living beings, from the most humble to the most complex, are related. The relationship is closer than we ever thought." Second, "Genetic engineering brought about a total change in the biological landscape as well as in the means of investigating it. Where it had been possible only to observe the surface of phenomena, it now became feasible to intervene in the heart of things."[46] Of course, Jacob's tropes—"landscape" and the "heart of things"—are archaic. As he is an "old European," to use a phrase from Habermas, we can be tolerant of Jacob's figurations; and since he is a wise European, we should be attentive to what he sees. But we should also be alert to the fact that our practices may well be outrunning our core metaphors.[47] In that case, inventiveness in the cultural sciences would have to be placed extremely high on an agenda of value orientations.

Consolations

Let us return to *Civilization and Its Discontents*. Freud concluded his book in a clinical manner, simultaneously incisive and hesitant. "The fateful question for the human species" is whether their civilization can master "the human instinct of aggression and self-destruction." But any answer to this question is unfortunately directly linked to the advance of knowledge; the remedy and the malady proceed together. "Men have gained control over the forces of nature to such an extent that with their help they would have no difficulty in exterminating one another to the last man. They know this, and hence comes a large part of their current unrest, their unhappiness and their mood of anxiety."[48] And indeed, the decades after 1930, when these sentences were written, would witness unparalleled slaughter and brutality in world history. Although Freud had offered his audience a predominantly pessimistic diagnosis, his tone should not, he says, be read as advocating any specific value judgments. "My impartiality," he added, "is made all the easier to me by my knowing very little about these things." However, what Freud does know "for certain . . . is that man's judgments of value follow directly his wishes for happiness—that, accordingly, they are an attempt to support his illusions with arguments. . . . I can offer them no consolation: for at bottom that is what they are all demanding—the wildest revolutionaries no less passionately than the most virtuous believers."[49] Freud was surely correct in foreseeing a prosperous, if discontented, future for the hard-working humans devoted to crafting themselves as prosthetic gods.

Freud's use of the term "consolation" [*trost*] is striking and unexpected. It is unexpected because clearly the would-be prosthetic gods are seeking happiness; hence they will not even notice that Freud isn't offering either happiness or consolation. The gift of consolation appears, rather, to be precisely what Freud can offer to himself and to those who would join him in his heroic *Wissenschaft*. To those, that is, who would bear the lack of solidarity that Weber posited as the price scientists pay for progress.

Consolation, however, need not be so bitter, and in English it falls on the sweeter end of a spectrum of physiognomy. "Consolation" is semantically layered. In English the transitive verb "to console" means to "alleviate the grief, sense of loss or trouble." The *Oxford English Dictionary* claims that the verb is modern. Its core meaning is "to support," for the verb is a transformation of the noun "console," first used in 1664 to refer to "an architectural member projecting from a wall to form a bracket for ornamentation." Although Freud disdained support for those seeking a firm stand for their ornamentation, he did hope to alleviate to some extent the sense of trouble of those seeking an orientation in life, especially a life in science, understood as the pursuit of enlightenment. Even so, enlightenment was a hard road, reserved, in Freud's view, only for the few strong enough to travel it; it was no longer a wave of beneficial historical progress, carrying along the many in its wake.

These German men sought to be upright within a modernity their scientific understanding had led them to see as yielding many dangers and few consolations. Their diagnosis is ever so close to Nietzsche's: humans would rather value something than nothing; an active nihilism is better than a reactive one. In that light, Freud's pathos and Weber's bathos can be seen as both courageous and virtuous.

INTRODUCTION: CONTEMPORARY EQUIPMENT FOR ANTHROPOLOGICAL
PROBLEMS OF MODERN SCIENCES
ANTHONY STAVRIANAKIS, GAYMON BENNETT, AND LYLE FEARNLEY

1. Mario Biagioli, ed., *The Science Studies Reader* (New York: Routledge, 1999); Edward J. Hackett et al., *The Handbook of Science and Technology Studies* (Cambridge, Mass.: The MIT Press, 2008).

2. Michel Foucault, "Polemics, Politics, and Problematizations: An Interview with Michel Foucault," in *The Foucault Reader*, ed. Paul Rabinow (New York: Pantheon, 1984), 111–119.

3. Paul Rabinow, *Anthropos Today: Reflections on Modern Equipment* (Princeton, N.J.: Princeton University Press, 2003), 6.

4. Hans Blumenberg, "'Imitation of Nature': Toward a Prehistory of the Idea of the Creative Being," *Qui Parle* 12, no. 1 (Spring/Summer 2000): 17–54.

5. Paul Rabinow, "Humanism as Nihilism: The Bracketing of Truth and Seriousness in American Cultural Anthropology," in *Social Science as Moral Inquiry*, ed. N. Haan, R. M. Bellah, P. Rabinow, and W. M. Sullivan (New York: Columbia University Press), 52–75; reprinted in *The Accompaniment* (Chicago: University of Chicago Press, 2011). Tobias Rees, "As If 'Theory' Is the Only Form of Thinking, and 'Social Theory' the Only Form of Critique: Thoughts on an Anthropology BST (Beyond Society and Theory)," *Dialectical Anthropology* 35 (2011): 341–365.

6. For example: Bruno Latour and Steve Woolgar, *Laboratory Life: The Construction of Scientific Facts* (Princeton, N.J.: Princeton University Press, 1986); Marilyn Strathern, *After Nature: English Kinship in the Twentieth Century* (Cambridge: Cambridge University Press, 1992); Donna Haraway, *Primate Visions: Gender, Race, and Nature in the World of Modern Science* (New York: Routledge, 1989).

7. Let us highlight, as an exemplar, the works of Jeanne Favret-Saada, whose studies of beliefs as practices, principally in her ethnographic and sociohistori-

cal accounts of different forms of accusation (such as witchcraft accusations in France and blasphemy accusations), are an alternative manner in which the rationalities of the moderns have been investigated. Jeanne Favret-Saada, "Relations de dépendance et manipulation de la violence en Kabylie," *L'Homme* 8, no. 4 (1968): 18–44; *Les mots, la mort, les sorts: la sorcellerie dans le bocage* (Paris: Gallimard, 1977); "Rushdie et compagnie: préalables à une anthropologie du blasphème," *Ethnologie française* 22, no. 3 (1992): 251–260; *Le christianisme et ses juifs: 1800–2000* (Paris, Le Seuil, 2004), with Josée Contreras; *Comment produire une crise mondiale avec douze petits dessins* (Paris, Les Prairies ordinaires, 2007); *Désorceler* (Paris: L'Olivier, 2009). Cf. Louis Dumont, *Essays on Individualism: Modern Ideology in Anthropological Perspective* (Chicago: University of Chicago Press, 1986); *German Ideology: From France to Germany and Back* (Chicago: University of Chicago Press, 1994).

8. On anthropology during this period of upheaval in divided Germany (1960–1989), see Andre Gingrich, "Anthropology in Four German-Speaking Countries: Key Elements of Post–World War II Developments to 1989," in Fredrik Barth, Andre Gingrich, Robert Parkin, Sydel Silverman, *One Discipline, Four Ways: British, German, French, and American Anthropology* (Chicago: University of Chicago Press, 2005).

9. Clifford Geertz, "An Inconstant Profession," in *Life Among the Anthros and Other Essays* (Princeton, N.J.: Princeton University Press, 2012), 187.

10. Clifford Geertz, "Ritual and Social Change: A Javanese Example," *American Anthropologist* 59, no. 1 (February 1957); *Religion of Java* (Chicago: University of Chicago Press); Hildred Geertz, *Modjokuto: Town and Village Life in Java* (Cambridge, Mass.: Center for International Studies, MIT, 1955).

11. Talcott Parsons and Edward A. Shils, eds., *Towards a General Theory of Action* (Cambridge, Mass.: Harvard University Press, 1951). Robert N. Bellah, *Tokugawa Religion* (Glencoe, Ill.: Free Press, 1957). Clifford Geertz, *Religion of Java* (Glencoe, Ill.: Free Press, 1960); *Peddlers and Princes: Social Change and Economic Modernization in Two Indonesian Towns* (Chicago: University of Chicago Press, 1963).

12. See, for example, Joseph G. Jorgensen and Eric R. Wolf, "A Special Supplement: Anthropology on the Warpath in Thailand," *New York Review of Books* 15, no. 9 (November 19, 1970): 1–4.

13. Dell H. Hymes, ed., *Reinventing Anthropology* (New York: Pantheon, 1969).

14. Laura Nader, "Up the Anthropologist: Perspectives Gained from Studying Up," in *Reinventing Anthropology*, ed. Dell H. Hymes (New York: Pantheon, 1969); Talal Asad, ed., *Anthropology and the Colonial Encounter* (Ithaca Press, 1973); Gayle Rubin, "The Traffic in Women: Notes on the 'Political Economy' of Sex," in *Toward an Anthropology of Women*, Rayna R. Reiter, ed. (New York: Monthly Review Press, 1975).

15. Philippe Ariès, *The Hour of Our Death* (Oxford: Oxford University Press, 1991). Marc Léopold Benjamin Bloch, *The Royal Touch: Sacred Monarchy and Scrofula in England and France* (Montreal: McGill-Queen's University Press, 1973). Lucien Febvre, *The Problem of Unbelief in the Sixteenth Century: The Religion of Rabelais* (Cambridge, Mass.: Harvard University Press, 1985); Jacques Le Goff, *La naissance du purgatoire* (Paris: Gallimard, 1981); Emmanuel Le Roy Ladurie, *Montaillou, village occitan* (Paris: Gallimard, 1975).

16. Lucien Lévy-Bruhl, *La mentalité primitive* (Alcan, 1922).

17. Emile Durkheim, "Individual and Collective Representations" (1898), in *Sociology and Philosophy*, trans. D. F. Pocock (London: Cohen and West, 1953).

18. Philippe Ariès, Paul Veyne, and Georges Duby, eds., *A History of Private Life: From Pagan Rome to Byzantium* (Cambridge, Mass.: Harvard University Press, 1992); Georges Duby, ed., *A History of Private Life II: Revelations of the Medieval World* (Cambridge, Mass.: Belknap Press of Harvard University Press, 1993); Roger Chartier, Phillippe Ariès, and Georges Duby, eds., *A History of Private Life III: Passions of the Renaissance* (Cambridge, Mass.: Harvard University Press, 1989); Michelle Perrot, Philippe Ariès, and Georges Duby, eds., *A History of Private Life IV: From the Fires of Revolution to the Great War* (Cambridge, Mass.: Harvard University Press, 1994); Philippe Ariès, Antoine Prost, Georges Duby, and Gérard Vincent, *A History of Private Life V: Riddles of Identity in Modern Times* (Cambridge, Mass.: Harvard University Press, 1991).

19. Eric. R. Wolf, *Europe and the Peoples Without History* (Berkeley: University of California Press, 1982).

20. Gyorgy Markus, "Why Is There No Hermeneutics of Natural Sciences? Some Preliminary Theses," *Science in Context* 1 (1987): 5–51.

21. George Marcus, "Ethnography Two Decades After Writing Culture: From the Experimental to the Baroque," *Anthropological Quarterly* 80, no. 4 (Fall 2007).

22. James Faubion, "History in Anthropology," *Annual Review of Anthropology* 22 (1993): 35–54.

23. Arjun Appadurai, *Modernity at Large: Cultural Dimensions of Globalization* (Minnesota: University of Minnesota Press, 1996); Nicholas B. Dirks, *The Hollow Crown: Ethnohistory of an Indian Kingdom* (Ann Arbor: University of Michigan Press, 1993).

24. George E. Marcus and James Clifford, eds., *Writing Culture: The Poetics and Politics of Ethnography* (Berkeley: University of California Press, 1986).

25. Marc Manganaro, ed., *Modernist Anthropology: From Fieldwork to Text* (Princeton, N.J.: Princeton University Press, 1990).

26. See Marilyn Strathern, "Out of Context: The Persuasive Fictions of Anthropology," *Current Anthropology* 28, no. 3 (June 1, 1987): 251–281. See

also the special issue of *Cultural Anthropology* 3, no. 4 (November 1988), devoted to inquiries into modernity. George Marcus and Michael M. J. Fischer, *Anthropology as Cultural Critique* (Chicago: University of Chicago Press, 1986).

27. See Massimo Mazzotti, "Introduction," in *Knowledge as Social Order: Rethinking the Sociology of Barry Barnes*, ed. Massimo Mazzotti (Hampshire: Ashgate, 2008); C. P. Snow, *The Two Cultures*, ed. Stefan Collini (Cambridge: Cambridge University Press, 1993).

28. David Bloor, *Knowledge as Social Imagery* (Chicago: University of Chicago Press, 1976).

29. Cf. Andrew Pickering, ed., *Science as Practice and Culture* (Chicago: University of Chicago Press, 1991).

30. Latour and Woolgar, *Laboratory Life*.

31. Bruno Latour, *The Pasteurization of France* (Cambridge, Mass.: Harvard University Press, 1993).

32. See Marilyn Strathern, "No Nature, No Culture: The Hagen Case," in *Nature, Culture, and Gender*, ed. Carol P. MacCormack and Marilyn Strathern (Cambridge: Cambridge University Press, 1980).

33. Donna Haraway, "A Cyborg Manifesto: Science, Technology, and Socialist-Feminism in the Late Twentieth Century," in *Simians, Cyborgs, and Women: The Reinvention of Nature* (New York: Routledge, 1991); *When Species Meet* (Minnesota: University of Minnesota Press, 2008).

34. Donna Haraway, "The Science Question in Feminism and the Privilege of Partial Perspective," *Feminist Studies* 14, no. 3 (Autumn, 1988): 575–599, 593.

35. Ibid., 594.

36. Marilyn Strathern, *The Gender of the Gift* (Berkeley: University of California Press, 1988), 9.

37. Max Weber, "Objectivity in Social Science," in *The Methodology of the Social Sciences*, trans. and ed. Edward Shils and Henry Finch (New York: Free Press, 1949), 67.

38. Maurice Bloch, "Language, Anthropology, and Cognitive Science," *Man* 26, no. 2 (June 1991): 183–198. Harvey Whitehouse, "Towards an Integration of Ethnography, History, and the Cognitive Science of Religion," in Harvey Whitehouse and James Laidlaw, *Religion, Anthropology, and Cognitive Science* (Durham, N.C.: Carolina Academic Press, 2007), 247–280.

39. Michel Foucault, *The Hermeneutics of the Subject* (New York: Palgrave Macmillan, 2005), 17.

40. Pierre Hadot and Arnold Ira Davidson, *Philosophy as a Way of Life: Spiritual Exercises from Socrates to Foucault* (New York: Wiley, 1995); Pierre Hadot, *The Inner Citadel: The Meditations of Marcus Aurelius* (Cambridge, Mass.: Harvard University Press, 1998). Jean-Pierre Vernant, *The Origins of Greek*

Thought (Ithaca, N.Y.: Cornell University Press, 1982); *Mortals and Immortals* (Princeton, N.J.: Princeton University Press, 1991).

41. Foucault, *Hermeneutics of the Subject*, 15.

42. Ibid.

43. As we will later indicate, one of the classic philosophical articulations of the problem of science and modernity as "research" is Martin Heidegger's essay "The Age of the World Picture" (1938). See also Hans-Jörg Rheinberger, *On Historicizing Epistemology*, trans. David Fernbach (Stanford, Calif.: Stanford University Press, 2010).

44. Roughly from the founding of the University of Berlin in 1808 to the U.S. Morrill Acts of 1862 and 1890. See further discussion below.

45. James Schmidt, "Introduction," in *What Is Enlightenment? Eighteenth-Century Answers and Twentieth-Century Questions*, ed. James Schmidt (Berkeley: University of California Press, 1996), 2.

46. Ibid., 3, citing Johann Karl Möhsen, "What Is to Be Done Toward the Enlightenment of the Citizenry? (1783)," also in ibid., 52.

47. Michel Foucault, *The Government of Self and Others, Lectures at the Collège de France, 1982–1983* (New York: Picador, 2010), 35.

48. Ibid, 35–36.

49. Foucault, "What Is Enlightenment?" Emphasis added.

50. Ibid., 116.

51. Hans-Jörg Rheinberger, "On the Historicity of Scientific Knowledge: Ludwik Fleck, Gaston Bacherland, Edmund Husserl," in David Hyder and Hans-Jörg Rheinberger, *Science and the Life-World: Essays on Husserl's Crisis of European Sciences* (Stanford, Calif.: Stanford University Press, 2009).

52. Ibid.

53. Ibid.

54. Max Weber, "Science as a Vocation," in *From Max Weber: Essays in Sociology*, ed. Hans Gerth and C. Wright Mills (Oxford: Oxford University Press, 1946), 138.

55. 7 U.S.C. § 304.

56. Laurence R. Veysey, *The Emergence of the American University* (Chicago: University of Chicago Press, 1970), 40–49.

57. Richard Hofstadter and Walter P. Metzger, *The Development of Academic Freedom in the United States* (New York: Columbia University Press, 1955).

58. Veysey, *The Emergence of the American University*.

59. Ibid., 79.

60. John Dewey, "Academic Freedom," *Educational Review* 23 (1902). Emphasis added, quoted in ibid., 346.

61. Veysey, *The Emergence of the American University*, 81.

62. John Dewey, *Essays on Experimental Logic* (Chicago: University of Chicago Press, 1916), 12.

63. John Dewey, *Reconstruction in Philosophy*, enlarged ed. (Boston: Beacon, 1948), xiii.

64. John Dewey, "Propositions, Warranted Assertibility, and Truth," *Journal of Philosophy* 38, no. 7 (March 27, 1941): 169–186, 170.

65. Dewey, *Reconstruction in Philosophy*, xxvii.

66. Fritz K. Ringer, *The Decline of the German Mandarins: The German Academic Community, 1890–1933* (Cambridge, Mass.: Harvard University Press, 1969), 3.

67. Weber, "Science as a Vocation," 139.

68. Ibid.

69. Ibid., 156.

70. "As asceticism began to change the world and endeavored to exercise its influence over it, the outward goods of this world gained increasing and finally inescapable power over men, as never before in history. Today its spirit has fled from this shell—whether for all time, who knows? Certainly, victorious capitalism has no further need for this support now that it rests on the foundation of the machine. Even the optimistic mood of its laughing heir, the Enlightenment, seems destined to fade away, and the idea of the 'duty in a calling' haunts our lives like the ghost of once-held religious beliefs." Max Weber, "The Protestant Ethic and the 'Spirit' of Capitalism," in *The Protestant Ethic and the "Spirit" of Capitalism and Other Writings*, ed. Peter Baehr and Gordon C. Wells (New York: Penguin, 2002), 121.

71. Weber, "Science as a Vocation," 156. As Weber said, "This, however, is plain and simple, if each finds and obeys the daemon who holds the fibers of his very life."

72. Max Weber, "'Objectivity' in Social Science and Social Policy," in *The Methodology of the Social Sciences*, trans. and ed. Edward Shils and Henry Finch (New York: Free Press, 1949), 72.

73. Ibid.

74. Ibid., 81.

75. Ibid., 84.

76. Ibid., 58.

77. Max Weber, *The Protestant Ethic and the Spirit of Capitalism* (New York: Routledge, 2000), 181. Cf. Adrian Wilding, "Max Weber and the 'Faustian Universality of Man,'" *Journal of Classical Sociology* 8 (2008): 67–88.

78. Weber, *The Protestant Ethic and the Spirit of Capitalism*, 181.

79. Edmund Husserl, *The Crisis of European Sciences and Transcendental Phenomenology: An Introduction to Phenomenological Philosophy* (Evanston, Ill.: Northwestern University Press, 1970), 14–15.

80. Ibid.

81. Ibid., 5–6.

82. Michel Foucault, "Life: Experience and Science," in *Aesthetics, Method, and Epistemology: Essential Works of Foucault 1954–1984*, ed. James D. Faubion (New York: New Press, 1998), 2:471.

83. Martin Heidegger, "The Age of the World Picture," in *The Question Concerning Technology, and Other Essays* (New York: Garland, 1977), 134–135.

84. Michel Foucault, "What Is Enlightenment?" in *The Foucault Reader*, ed. Paul Rabinow (New York: Pantheon, 1984).

85. Robert Savage, "Translator's Afterword," in *Paradigms for a Metaphorology* (Ithaca, N.Y.: Cornell University Press, 2010), 140.

86. Ibid., 141.

87. Ibid.

88. Hans Blumenberg, *The Legitimacy of the Modern Age* (Cambridge, Mass.: MIT Press, 1988), 229.

89. Gaston Bachelard, *The New Scientific Spirit* (Boston: Beacon, 1984); Jean Cavaillès and Gaston Bachelard, *Sur la logique et la théorie de la science* (Paris: Vrin, 1997); Georges Canguilhem, *Ideology and Rationality in the History of the Life Sciences* (Cambridge, Mass.: MIT Press, 1988); Michel Foucault, *History of Madness* (New York: Routledge, 2006); Max Scheler, *Probleme einer Soziologie des Wissens* (Duncker & Humblot, 1924). Karl Mannheim, *Ideology and Utopia: An Introduction to the Sociology of Knowledge* (Psychology Press, 1936); Ludwig Fleck, *Genesis and Development of a Scientific Fact* (Chicago: University of Chicago Press, 1981).

90. Bachelard, *The New Scientific Spirit*.

91. Georges Canguilhem, *A Vital Rationalist: Selected Writings from Georges Canguilhem* (New York: Zone, 2000).

92. Georges Canguilhem, *"The Living and Its Milieu" in Knowledge of Life* (New York: Fordham University Press), 113.

93. Ibid., 118.

94. Ibid., 120.

95. "À travers l'analyse d'expériences historiques, collectives, sociales, liées à des contextes historiques précis, comment peut-on faire l'histoire d'un savoir, l'histoire de nos connaissances et comment des objets nouveaux peuvent-ils arriver dans le domaine de la connaissance, peuvent-ils se présenter comme objets à connaitre." André Berten and Michel Foucault, "Entretien avec Michel Foucault," *Les Cahiers du GRIF* 37–38 (1988): 11. Our translation.

96. Michel Foucault, "On the Genealogy of Ethics: An Overview of Work in Progress," in *The Foucault Reader*, ed. Paul Rabinow (New York: Pantheon, 1984). Michel Foucault, *The History of Sexuality*, vol. 2: *The Use of Pleasure* (New York: Pantheon, 1985), 32.

97. James Faubion's extensions of Foucault's parameters to include those of recruitment, ethical valuation, justification, and judgment are important because they indicate the structure and organization of the relative complexity (or simplicity, as the case may be) of the domain in which subject positions are occupied, and the work, ends, substance, and judgments that they require. The point here is that within a complex domain, of which contemporary sciences are no exception, "the occupant of one or another ethically marked subject

position finds himself or herself or itself to be yet a second ethical subject."
James D. Faubion, *An Anthropology of Ethics* (Cambridge: Cambridge University Press, 2011), 14.

98. Foucault invents the term *alēthurgia* from the Greek word *alēthourgēs*—
"acting truly"—to refer to "the set of possible procedures, verbal or otherwise,
by which one brings to light what is posited as true, as opposed to the false, the
hidden, the unspeakable, the unforeseeable, or the forgotten. We could call
'alethurgy' that set of procedures and say that there is no exercisee of power
without something like an alethurgy." Michel Foucault, *The Courage of Truth:
The Government of Self and Others II; Lectures at the Collège de France, 1983–1984*
(New York: Picador, 2012), 20.

99. Ibid., 3–4.

100. Foucault, *The Hermeneutics of the Subject*, 19.

101. Paul Rabinow, *Marking Time, on the Anthropology of the Contemporary*
(Princeton, N.J.: Princeton University Press, 2008), 2.

AN ANSWER TO THE QUESTION: "WHAT IS ENLIGHTENMENT?"
IMMANUEL KANT

1. Literal translation: "Dare to be wise," Horace, *Epodes* I, 2, 40. Cf.
Friedrich Schiller, *On the Aesthetic Education of Man*, ed. and trans. Elizabeth
M. Wilkinson and L. A. Willoughby (Oxford, 1967), 74ff; cf. Franco Venturi,
"Was ist Aufklärung? Sapere Aude!" *Revista Storica Italiana* 71 (1959): 119ff.
Venturi traces the use made of this quotation from Horace throughout the
centuries.

2. "Those who have come of age by virtue of nature."

3. Frederick II (The Great) King of Prussia (1749–1786).

4. "Caesar is not above the grammarians."

5. Frederick II (The Great) King of Prussia (1749–1786).

6. This allusion amounts to a repudiation of Julien Offray de Lamettrie's
(1709–1751) materialism as expressed in *L'homme machine* (1748).

7. Anton Friedrich Büsching (1724–1793), professor in the University of
Göttingen, theologian, and leading geographer of the day, editor of *Wochentliche Nachrichten von neuen Landkarten, geographischen, statistischen, une historischen Buchern*. Kant's reference is to volume 12, 1784 (Berlin 1785), 291.

8. Moses Mendelssohn (1729–1786), a leading philosopher of the German
Enlightenment. The reference is to Mendelssohn's essay "Über die Frage; was
heist Aufklärung?" *Berlinische Monatschrift* 4 (September 9, 1784): 193–200.

SCIENCE AS A VOCATION
MAX WEBER

1. The German word *Beruf* has a workaday meaning of "profession" but,
rooted as it is in *rufen*, "to call," has strong overtones of "vocation" or "call-

ing." Both meanings are active in Weber's usage, and each has been used here where it seemed appropriate. The term *Wissenschaft* means "science" but can refer to any academic discipline or body of knowledge. Thus not only the social sciences but even literary studies, musicology, or linguistics are all called *Wissenschaft*. We have kept "science" here, even though it may seem strange to the English reader who is accustomed to using it with reference to the natural sciences. But we have also used "scholarship" or "studies" and the adjective "academic" where English usage required it.

2. This refers to the German Habilitation, a second doctorate by dissertation that is usually taken about ten years after the Ph.D. and serves as the springboard to an academic career.

3. German students used to have a *Studienbuch*, a notebook in which they registered the courses they were taking in their field. They then had to pay a fixed fee for each course. For staff on a full salary—that is, professors—these tuition fees were a welcome extra. For the unsalaried *Privatdozent*, these fees were the sole source of income.

4. Ludwig Uhland (1787–1862) was a romantic poet who made his name with ballads and poems in a folk style. He also wrote political poetry with a strongly patriotic emphasis. He was always in the second rank and, while still famous in Weber's day, he is now largely neglected, surviving chiefly in school anthologies.

5. Weber used the English word.

6. Weber was made a full professor in what was then known as political economy (a social science that focused on the state and its resources) at the University of Freiburg in 1895, when he was only thirty-one.

7. In Germany professors are civil servants and are still appointed by a procedure in which the faculties submit a shortlist of names to the Ministry of Education, which then makes the final choice.

8. Hermann Helmholtz (1821–1894) was one of the outstanding German scientists of the nineteenth century, notable for his contributions in both physics and physiology. His achievements include the formulation of the principle of the conservation of energy. Leopold von Ranke (1795–1886) was a leading German historian whose search for historical objectivity greatly influenced historiography throughout Europe. Both had chairs in Berlin.

9. *Lasciate ogni speranza* [*voi ch'entrate*]*!* (Abandon all hope, [ye who enter here]!), Dante, *Inferno*, canto 3, line 9. This is the inscription on the lintel above the gates of Hell.

10. Robert Mayer (1814–1878) was a German doctor who made his name following his observation that in the tropics the color difference between venous and arterial blood was smaller than in temperate climates. He inferred that the higher temperatures made it unnecessary to convert as much food in order to conserve body heat as in colder latitudes. This led him to develop an influential theory of the equivalence of heat and physical labor.

11. Rudolph von Ihering (1818–1892), jurist and professor at Göttingen from 1872 on.

12. Karl Weierstrass (1815–1897). He is regarded as one of the founding fathers of modern functional analysis.

13. For example, in *Phaedrus* 245, where Plato writes, "If a man comes to the door of poetry untouched by the madness of the Muses, believing that technique alone will make him a good poet, he and his sane compositions never reach perfection but are utterly eclipsed by performances of the inspired madman."

14. Genesis 25:8.

15. Weber evidently has such works as *The Death of Ivan Ilyich* (1886) and *Resurrection* (1899) in mind.

16. That is, the level of a university graduate with a doctorate.

17. Jan Swammerdam (1637–1680) was a Dutch naturalist who undertook pioneering studies with the microscope. Among other discoveries, he was the first to observe and describe red blood cells (1658). The quotation here is taken from his *Algemeene Verhandeling van bloedeloose diertjens* (1658) (*The Natural History of Insects*, 1792).

18. Philip Jakob Spener (1635–1705) was a leading figure of German Pietism. This movement initiated a spiritual renewal of Protestantism through an emphasis on personal improvement and upright conduct, which it held to be the most important manifestations of the Christian faith. It had a profound influence on German religious thought and, more generally, on German literature and culture.

19. "I tell you: one must have chaos in one, to give birth to a dancing star.... Alas! The time is coming when man will give birth to no more stars. Alas! The time of the most contemptible man is coming, the man who can no longer despise himself. Behold! I shall show you the *Last Man*. 'What is love? What is creation? What is longing? What is a star?' Thus asks the Last Man and blinks.... 'We have discovered Happiness,' say the Last Men and blink." See Nietzsche's *Thus Spoke Zarathustra*, trans. R. J. Hollingdale (Harmondsworth: Penguin, 1969), 46. Hollingdale prefers "the Ultimate Man."

20. It has not been possible to find the definitive source of this quotation. The statement may be derived from Leo Tolstoy, "What Should We Do Then?" in *Collected Works*, trans. Leo Weiner (New York: AMS, 1968), 17:249–289 (chapters 32–37). See note above. More of Tolstoy's criticism of science can be found in Leo Tolstoy, *A Confession and What I Believe*, trans. Aylmer Mande (Oxford: Oxford University Press; London: Humphrey Milford, 1938). In Chapter 5 he describes how he is "brought to the verge of suicide" by his inability to discover whether there "is any meaning in my life that the inevitable death awaiting me does not destroy." And he concludes a lengthy discussion with the assertion that science in all its forms is unable to disclose such a meaning (26–35).

21. Dietrich Schäfer (1845–1929) was a historian who taught at Jena, Breslau, Tübingen, and Heidelberg, as well as Berlin. He was a member of the Pan-German Society, and his nationalist, annexationist views became increasingly strident during World War I. He also advocated the unrestricted use of submarine warfare.

22. Friedrich Wilhelm Foerster (1869–1966) was an educationist and politician who held chairs in Vienna and Munich. His strongly Christian and pacifist views led him to be highly critical of Prussian and German policies during the nineteenth and twentieth centuries. His pacifist views led to a year's suspension from his post at Munich University in 1916. His reinstatement in 1917 was followed by violent clashes between left-wing and right-wing students. After the war he emigrated to Switzerland.

23. Jeremiah 2:2.

24. In Isaiah 53 we find, inter alia: "To whom hath the arm of the Lord been revealed? For he grew up before him as a tender plant, and as a root out of a dry ground; he hath no form nor comeliness; and when we see him, there is no beauty that we should desire him. He was despised, and rejected of men; a man of sorrows, and acquainted with grief; and as one from whom men hide their face he was despised, and we esteemed him not." Psalm 22 (not 21 as in Weber) contains a similar evocation of a man despised and abandoned by God ("My God, my God, why hast thou forsaken me?") but whose faith is intact.

25. Matthew 5:39.

26. Luke 10:42.

27. J. W. von Goethe, *Faust*, part 2, trans. Philip Wayne (Harmondsworth: Penguin, 1959), p. 99, ll. 6817–6818.

28. This quotation has not been identified, but see, for example, "How Is Natural Science Possible?" in Immanuel Kant, *The Critique of Pure Reason*, trans. Paul Guyer and Paul W. Wood (Cambridge: Cambridge University Press, 1997), 147.

29. Georg von Lukács (1885–1971) became a leading Marxist philosopher at the end of World War I. Before that he was a noted literary critic and philosopher of art, associated with a circle around Max Weber. He published two influential books on literature, *Die Seele und die Formen* (1909) (*Soul and Form* [Cambridge, Mass.: MIT Press, 1974]) and *Theorie des Romans* (1916) (*The Theory of the Novel* [Cambridge, Mass.: MIT Press, 1971]).

30. "I believe not what [is absurd], but because it is absurd." Generally attributed now to Tertullian (c. 155/160–after 220] rather than St. Augustine.

31. Isaiah 21:11–12. The translation given in the text is a direct translation from Martin Luther's German, of which Weber's text gives a slight paraphrase. This diverges from the traditional English renderings, which arguably may puzzle the lay reader and fail to make Weber's reason for quoting it clear.

Thus, the Revised Version has: "The watchman said, The morning cometh, and also the night: if ye will inquire, inquire ye: turn ye, come."

32. The quotation is from Goethe, *Wilhelm Meisters Wanderjahre*, which contains the exchange, "What is your duty? The challenge of the day." *Weimarer Ausgabe* (Weimar, 1907), vol. 42, section 2, p. 187.

33. Weber uses the word *Dämon*, which means both "daemon" and "demon." A "daemon" is an inner or attendant spirit. The term goes back at least to Socrates in the *Symposium*, but it was given currency among the educated German public by a poem by Goethe with the title *Dämon*, which was obviously known to Weber and contains inter alia the lines: "Even as the sun and planets stood, to salute one another on the day you entered the world—even so you began straightaway to grow and have continued to do so, according to the law that prevailed over your beginning. It is thus that you must be, you cannot escape yourself. . . ."

RECONSTRUCTION AS SEEN TWENTY-FIVE YEARS LATER
JOHN DEWEY

1. The obvious insufficiency of psychological theories on this point has played a part in developing the formalisms already noted. Instead of using this insufficiency as ground for reconstruction of the psychological theory, the defective view was accepted qua psychology and hence was used as a ground for a "logical" theory of knowing that shut out entirely all reference to the factual ways in which knowledge advances.

2. C. D. Darlington, *Conway Memorial Lecture on The Conflict of Society and Science* (London: Watts & Co., 1948); italics not in text.

3. It is well worth recalling that for quite a while Newton ranked as "philosopher" of the division of that subject still classified as "natural" in distinction from metaphysical and moral. Even by his followers his deviations from Descartes were treated as matter not of physical science but of "natural philosophy."

WHAT IS ENLIGHTENMENT?
MICHEL FOUCAULT

1. Giambattista Vico, *The New Science*, 3rd ed. (1744), abridged and trans. T. G. Bergin and M. H. Fisch (Ithaca, N.Y.: Cornell University Press, 1970), 370, 372.

2. Charles Baudelaire, *The Painter of Modern Life*, trans. Jonathan Mayne (London: Phaidon, 1964), 13.

3. Charles Baudelaire, "On the Heroism of Modern Life," in *The Mirror of Art*, trans. Jonathan Mayne (London: Phaidon, 1955), 127.

4. Baudelaire, *The Painter of Modern Life*, 12.

5. Ibid.

THE "TRIAL" OF THEORETICAL CURIOSITY

HANS BLUMENBERG

1. Victor Hugo, *William Shakespeare* (1864), part 1, book 3, section 4, "La science cherche le mouvement perpétuel. Elle l'a trouvé; c'est elle-même," trans. M. B. Anderson (Chicago: McClurg, 1911), 105.

2. Hans Lipps, "Die philosophischen Probleme der Naturwissenschaft" (1931), in *Die Wirklichkeit des Menschen* (Frankfurt: Klostermann, 1954), 193.

3. J. Chr. P. Erxleben, *Anfangsgründe der Naturlehre*, 4th ed. (Göttingen, 1787), section 15.

4. "'Reflektierte' Neugierde." Like our "reflect," *reflektieren* conveys the ideas of considering and of "reflecting on" something, but in addition it carries a clearer suggestion than our term does of "reflexiveness," of the possibility of considering or reflecting on one's own actions and inclinations *as* one's own, and thus of self-consciousness.

5. Jürgen Mittelstrass, "Bildung und Wissenschaft. Enzyklopädien in historischer und wissenssoziolgischer Betrachtung," *Die wissenschaftliche Redaktion* 4 (1967): 81–104.

6. Otto Liebmann, *Die Klimax der Theorien* (Strasbourg: 1884), 4–5.

7. Hermann Lübbe, "Die geschichtliche Bedeutung der Subjektivitätstheorie Edmund Husserls," *Neue Zeitschrift für systematische Theologie* 2 (1960): 319. Now in *Bewusstsein in Geschichten. Studien zur Phänomenologie der Subjektivität* (Freiburg: Rombach, 1972), 31–32.

8. A remark reportedly made by Archimedes to Roman soldiers who, after conquering his city, Syracuse, were on the point of killing him.

9. Mittelstrass, "Bildung und Wissenschaft," 83.

10. Montesquieu, *Discours sur les motifs qui doivent nous encourager aux sciences, Ouevres completes* (Paris: Didot), 579: "Le premier, c'est la satisfaction intérieure que l'on ressent lorsque l'on voit augmenter l'excellence de son être, et que l'on rend intelligent un être intelligent. Le second, c'est une certaine curiosité que tous les hommes ont, et qui n'a jamais été si raisonable que dans ce siècle-ci. Nous entendons dire tous les jours que les bornes des connaissances des hommes viennent de'être infiniment reculées, que les savants étonnés de se trouver si savants, et que la grandeur des succès les a fait quelquefois douter de la vérité de succès . . ."

11. Cicero, *Tusculanae Disputationes* V.10: "Socrates autem primus philosophiam devocavit e caelo et in urbibus conlocavit et in domus etiam introduxit et coëgit de vita et moribus rebusque bonis et malis quaerere."

12. Lactantius, *Divinae Institutiones* III.20.10. Erasmus included the sentence in his *Adagia* I.6.69 and interpreted it as follows: "Dictum Socraticum deterrens a curiosa investigatione rerum coelestium et arcanorum naturae." Characteristic of the attempt to trace the sentence back to its original context is Erasmus's conclusion that one could also interpret it thus: "Quae infra

nos, nihil ad nos, ubi significamus res leviusculas, quam ut nobis curae esse debeant."

JUSTIFICATIONS OF CURIOSITY AS PREPARATION FOR THE ENLIGHTENMENT
HANS BLUMENBERG

1. Nietzsche, *Beyond Good and Evil*, part 5, §188, in *The Basic Writings of Nietzsche*, trans. Walter Kaufmann (New York: Random House, 1968), p. 291.

2. Goethe, *Fragmente zu Wissenschaftsgeschichte, Werke*, ed. E. Beutler (Zurich: Artemis, 1948–), 17:757.

3. On "pedantry as the substitute attitude of a consciousness which is blocked from meeting its needs": T. W. Adorno, introduction to Emile Durkheim, *Soziologie und Philosophie* (Frankfurt: Suhrkamp Verlag, 1967), 32n; commenting on the passage corresponding to this one in *Die Legitimät der Neuzeit* (Frankfurt: Suhrkamp, 1966), 380–381.

4. Justus Liebig, "Francis Bacon von Verulam und die Geschichte der Naturwissenschaften" (1863), in *Reden und Abhandlungen* (Leipzig, 1874), 233–234.

5. Francis Bacon *Novum organum* II.9. On the distinction between nature's *cursus consuetus* and its *praeter-generationes*, see his *De augmentis scientiarum* II.2.

6. *Novum organum* praefatio: ". . . ut mens suo jure in rerum naturam uti possit."

7. *Novum organum* I.129: "Recuperet modo genus humanum jus suum in naturam quod ei ex dotatione divina competit, et detur ei copia: usum vero recta ratio et sana religio gubernabit."

8. *Novum organum* praefatio: "Id tamen posteris gratum esse solet, propter usum operis expeditum et inquisitionis novae taedium et impatientiam."

9. *Novum organum* praefatio: "Nemo enim rei alicuius naturam in ipsa re recte aut feliciter perscrutatur; verum post laboriosam experimentorum variationem non acquiescit, sed invenit quod ulterius quaerat." Hence the importance of "negative instances" in the cognitive process; man cannot know *ab initio contemplationis*; rather his path goes by the *"procedere primo per negativas . . . post omnimodam exclusionem"* (*Novum organum* II.15, 2).

10. *Novum organum* I.28: ". . . ex rebus admodum variis et multum distantibus sparsim collectae (sc. interpretationes) . . ."

11. *Novum organum* I.98: ". . . occulta naturae magis se produnt per vexationes artium . . ." Here it is presupposed that the Aristotelian distinction between "natural" and "violent" movement is no longer made or should no longer be made; otherwise the latter could not provide information that can be carried over to the former. An indirect reference to its object defines the "interpretation" of nature: ". . . omnis verior interpretatio naturae conficitur per instantias, et experimenta idonea et apposita; ubi sensus de experimento

tantum, experimentum de natura et re ipsa judicat"; *Novum organum* I.50). In contrast to this "translated" explanation of nature, the instrumental strengthenings of the senses, the *organa ad amplificandos sensus*, lose importance for Bacon.

12. F. Schalk, "Zur Vorgeschichte der Diderotschen Enzyklopädie," *Romanische Forschung* 70 (1958): 40ff: "The peculiar disdain, both of mathematics and of the mechanical aids to research, that Bacon exhibits and that brought on him the censure of his critics in the nineteenth (the 'technical') century becomes intelligible only when one turns one's gaze, with him, to the Adam who ruled the cosmos by giving names." But the typified nonviolent dominion by means of the word is suspended on this side of paradise and becomes a utopian figure; where Bacon describes paths to knowledge in the present, his language is filled with expressions of toil and violence. The "idea of the *regnum hominis* as the dominion of the magicians over the cosmos and the management of this dominion in the service of humankind" (46) gives its character only to the totality of completed knowledge.

13. Francis Bacon *Valerius terminus* 1: "In aspiring to the throne of power the angels transgressed and fell; in presuming to come within the oracle of knowledge man transgressed and fell."

14. *Valerius terminus* 1: ". . . he was fittest to be allured with appetite of light and liberty of knowledge; therefore this approaching and intruding into God's secrets and mysteries was rewarded with a further removing and estranging from God's presence."

15. *Valerius terminus* 1: ". . . it was not that pure light of natural knowledge, whereby man in Paradise was able to give unto every living creature a name according to his propriety, which gave occasion to the fall; but it was an aspiring desire to attain to that part of moral knowledge which defineth of good and evil, whereby to dispute God's commandments and not to depend upon the revelation of his will, which was the original temptation."

16. *Valerius terminus* 1: ". . . as if according to the innocent play of children the divine Majesty took delight to hide his works, to the end to have them found out . . ." Compare *Novum organum* I.132.

17. *Valerius terminus* 1: ". . . God hath framed the mind of man as a glass capable of the image of the universal world, joying to receive the signature thereof as the eye is of light, yet not only satisfied in beholding the variety of things and vicissitude of times, but raised also to find out and discern those ordinances and decrees which throughout all these changes are infallibly observed."

18. *Valerius terminus* 1: "And although the highest generality of motion or summary law of nature God should still reserve within his own curtain, yet many and noble are the inferior and secondary operations which are within man's sounding."

19. *Valerius terminus* 1: ". . . but it is a restitution and reinvesting (in great part) of man to the sovereignty and power (for whensoever he shall be able to

call the creatures by their true names he shall again command them) which he had in his first state of creation."

20. *Valerius terminus* 1: "And therefore knowledge that tendeth but to satisfaction is but as a courtesan, which is for pleasure and nor for fruit or generation." Compare *Valerius terminus* 9.

21. *Valerius terminus* 1: ". . . the new-found world of land was not greater addition to the ancient continent than there remaineth at this day a world of inventions and sciences unknown, having respect to those that are known, with this difference, that the ancient regions of knowledge will seem as barbarous compared with the new, as the new regions of people seem barbarous compared to many of the old." Compare *Valerius terminus* 5.

22. *Valerius terminus* 17: "That those that have been conversant in experience and observation have used, when they have intended to discover the cause of any effect, to fix their consideration narrowly and exactly upon that effect itself with all the circumstances thereof, and to vary the trial thereof as many ways as can be devised; which course amounteth but to a tedious curiosity, and ever breaketh off in wondering and not in knowing . . ."

23. *Novum organum* praefatio: "Quin illis hoc ferre solenne est, ut quicquid ars aliqua non attingat id ipsum ex eadem arte impossibile est statuant." Compare *Novum organum* I.88.

24. *Novum organum* I.109, I.122. *Valerius terminus* 11.

25. Leibniz to Johannes Bernoulli (February 21, 1699), in *Mathematische Schriften*, ed. C. I. Gerhardt (Berlin and Halle: 1850–1863), 3:574: "Scio multos dubitare, ut insinuas, an nos possimus cognoscere, quid sit Sapientiae Justitiaeque divinae conforme. Puto tamen, ut Geometria nostra et Arithmetica etiam apud Deum obtinent, ita generales boni justique leges, mathematicae certitudinis et apud Deum quoque validas esse." *Animadversiones in partem generalem Principiorum Cartesianorum*, ed. C. I. Gerhardt, in *Philosophische Schriften* (Berlin: 1875–1890), 4:375ff, commenting on II.45: ". . . sed natura, cujus sapientissimus Auctor pefectissimam Geometriam exercet, idem observat, alioqui nullus in ea progressus ordinatus servaretur." *Mathesis divina*, for its part, is not an independent and final principle, but rather is founded on the principle of sufficient reason; it is the form in which the rational explanation of realized possibilities displays itself: *Tentamen anagogicum*, in *Philosophische Schriften*, 7:273–274, 304; compare 2:105, 438; 3:51; 4:216; and 7:191. This converges with the observation that Leibniz's divine geometry is not spatial/intuitive, but rather, after the model of analytic geometry, the epitome of the generative calculus of bodies. Spatial/corporeal nature is only the pictorial equivalent of this geometry; but the Platonic sense of this assertion is suspended, though for Kepler it was still bound up with the God Who practices geometry: "Non aberrat . . . ab archetypo suo Creator, geometriae fons ipsissimus, et, ut Plato scripsit, aeternam exercens geometriam . . ." *Harmonice mundi*, in *Gesammelte*

Werke, ed. Caspar (Munich: Beck, 1937–), 6:299. The progress of mathematics represents the process by which man penetrates into the coherence of *ratio sufficiens* and the world calculus and thus at the same time withdraws his knowledge from the requirement of legitimation.

26. Galileo *Dialogo dei massimi sistemi* I: ". . . adunque bisognerá dire che né anco la natura abbia inteso il modo di fare un intelletto che intenda."

27. *Dialogo* I: ". . . dico che l'intelletto umano ne intende alcune cosí perfettamente, e ne ha cosí assoluta certezza, quanto se n'abbia l'istessa natura . . ."

28. *Dialogo* I: ". . . poiché arriva a comprenderne la necessitá, sopra la quale non par che possa esser sicurezza maggiore."

29. *Dialogo* I: ". . . anzi, quando io vo considerando quante e quanto maravigliose cose hanno intese investigate ed operate gli uomini, pur troppo chiaramente conosco io ed intendo, esser la mente umana opera di Dio, e delle piú eccellenti." The argument is less medieval than it looks; it assigns the burden of giving a satisfactory account of the intellect's author to its verifiable accomplishments instead of presupposing illumination and man's having been created in the image of God.

30. *Dialogo* IV: "Mirabile e veramente angelica dottrina: alla quale molto concordemente risponde quell'altra, pur divina, la quale, mentre ci concede il disputare intorno alla constituzione del mondo, ei soggiugne (forse acciò che l'esercizio delle menti umane non si tronchi o anneghittisca) che non siamo per ritrovare l'opera fabbricata dalle Sue mani."

31. *Materialien zu Brechts "Leben des Galilei"* (Frankfurt: Suhrkamp, 1963), 12–13. In addition, the note in "Construction of a Role" (60): "He appealed to his irresistible inquisitive drive, as a detected sex criminal might appeal to his glands." The anthropological systematics in which Brecht's figure of Galileo belongs are clarified by the categories of his theory of the theater; in place of the Aristotelian dyad of "pity and fear" in the dramatic reception, there enter, for the non-Aristotelian experimental theater, "curiosity and helpfulness" (compare *Materialien*, 163, 169).

32. Galileo *Discorsi* [*Discourses Concerning Two New Sciences*] I: "Ma se le digressioni possono arrecarci la cognizione di nuove vertiá, che pregiudica a noi, non obligati a un metodo serrato e conciso, ma che solo per proprio gusto facciamo i nosti congressi, digredir ora per non perder quelle notizie che forse, lasciata l'incontrata occasione, un'altra volta non ci si rappresenterebbe? anzi chi sa che bene spesso non si possano scoprir curiositá piú belle delle primariamente cercate conclusioni?"

33. Galileo *Dialogo* III: ". . . un conoscere che infinite cose restano in natura incognite."

34. Descartes to Mersenne (October 11, 1638), in *Oeuvres*, ed. Adam and Tannery (Paris: Cerf, 1897–1913), 2:380: "Il me semble qu'il manque beaucoup en ce qu'il fait continuellement des digressions, et ne s'areste point à

expliquer tout à fait une matière; ce qui monstre qu'il ne les a point examinées par ordre, et que, sans avoir considéré les premières causes de la nature, il a seulement cherché les raisons de quelques effets particuliers, et ainsi qu'il a basti sans fondement."

35. U. Ricken, *"Gelehrter" und "Wissenschaft" im Französischen. Beiträge zu ihrer Bezeichnungsgeschichte vom 12. bis 17. Jahrhundert* (Berlin: Akademie Verlag, 1961), 167–168. In the *Discours de la Methode* (I), the *sciences curieuses* are the disciplines lying apart from the Scholastic curriculum. See Etienne Gilson in Descartes, *Discours de la Méthode, texte et commentaire* (Paris: J. Vrin, 1947), 109, with the gloss reproduced there from Furetière's *Dictionnaire universel*. A trace of the magic and the mantic remains in such unusual interests, an excess over what is useful in life, which bars them from the system of the Method (Gilson, 120–121, 140–141). The antithesis *pour mon utilité/pour ma curiosité* is found in the letter to Mersenne of February 9, 1639 (*Oeuvres*, ed. Adam and Tannery, 2:499). The defense of the *curiosité* by now is only incidental and without argumentative effort: "Ce n'est pas un crime d'estre curieux de l'Anatomie . . . j'allois quasi tous les jours en la maison d'un boucher, pour luy voir tuer les bestes . . ." (2:621). The unanswerable questions automatically exclude themselves under the criterion of the Method because their treatment evades mathematization: *Regulae ad directionem ingenii* 8 (10:398). There remains radical significance of the carefulness exerted in assuring oneself of the evidence: "Atque haec omnia quo diutius et curiosius examino, tanto clarius et distinctius vera esse cognosco . . ." (*Meditationes* III.16; *Oeuvres*, 7:42).

36. Furetière, *Dictionnaire universel* (1690): "C'est un curieux de livres, de médailles, d'estampes, de tableaux, de fleurs, de coquilles, d'antiquités, de choses naturelles." The idleness of dilettantism is indicated by a commonplace remark: "C'est un chymiste curieux qui a fait de belles experiences, de belles descouvertes." A *Collegium Naturae Curiosorum* is put together in 1650; uncounted book titles offer "curious" objects for "curious" readers, like Caspar Schott's *Technica Curiosa* of 1664 and *Physica Curiosa* of 1662, W. H. von Hochberg's *Georgica Curiosa* of 1682/1687, and the *Schatzkamer rarer und neuer Curiositäten in den allerwunderbahrsten, Würckungen der Natur und Kunst* (Hambug: 1686). See H. Bausinger, "Aufklärung und Aberglaube," *Deutsche Vierteljahresschrift für Literaturwissenschaft und Geistesgeschichte* 37 (1963): 436–437.

37. Jakob Brucker, *Kurtze Fragen aus der Philosophischen Historie I* (Ulm, 1731), 223–224 (I 2, chap. 1, q. 2).

38. Fontenelle, *Histoire des Oracles*, ed. L Maigron (Paris: Cornély, 1908), 67. In the same year, 1686, in his influential *Entretiens sur la pluralité des mondes*, ed. A. Calame (Paris: Didier, 1966), 17, Fontenelle sees philosophy as a result of the paradoxical combination of curiosity and shortsightedness: "Toute la Philosophie . . . n'est fondée que sur deux choses, sur ce qu'on a l'esprit curieux et les yeux mauvais . . ."

39. J. G. Sulzer, *Philosophische Schriften II* (Leipzig: 1781), 114.

40. Brucker, *Kurtze Fragen aud der Philosophischen Historie II*, 880–883.

THE QUESTION OF NORMALITY IN THE HISTORY OF BIOLOGICAL THOUGHT
GEORGES CANGUILHEM

1. *Geschichte der biologischen Theorien in der Neuzeit*, vol. 1, part 2, rev. ed. (Leipzig and Berlin, 1913), preface, viii: "Auch von den Biologen wurde ein Galilei, ein Descartes als Begründer der neuen Auffassung des Lebens gepriesen, obwohl an diselben keine beachtenswertere biologische Idée anzuknüpfen ist."

2. Francois Jacob, *La logique du vivant* (Paris: Gallimard, 1970), 302.

3. Cf. E. Aziza-Shuster, *La médecin de soi-même* (Paris: PUF), chap. 1.

4. Ibid. My italics.

5. Salvador Luria, *Life: The Unfinished Experiment* (1973).

6. See my preface to the modern edition of Bernard's *Leçons sur les phénomènes de la vie communs aux animaux et aux végétaux* (Paris: Vrin, 1966), 207–218.

7. See my *Le normal et le pathologique* (Paris: PUF, 1966), 207–218.

8. Though normally I would resist the temptation to read old texts with today's eyes as anticipations of things to come, I cannot resist citing two passages from Cuvier's *Histoire des progrès des sciences naturelles de 1789 jusqu'à ce jour* (1810; Nelle ed., 1834): "Life is a constant turbulence, whose direction, however complex, remains constant, as does the species of molecules involved, though not the individual molecules themselves. On the contrary, the matter that presently constitutes the living body will soon cease to do so, yet it is the repository of the force that will constrain the future matter to move in the same direction. Thus the form of these bodies is more essential to them than their matter, for the latter constantly changes, while the former is preserved, and because it is forms that constitute the differences between species and not combinations of matter, which are practically the same in all" (187); and further, "One misconstrues the nature of life by thinking of it as a mere bond that holds together the elements of the living body. On the contrary, it is a spring that constantly moves and transports those elements" (210).

9. [Léon Brillouin, *Vie, matière, et observation* (Paris: Albin Michel, 1959), 105.—Trans.]

10. Marjorie Greene, *Approaches to a Philosophical Biology* (New York: Basic Books, 1965).

THE LIVING AND ITS MILIEU
GEORGES CANGUILHEM

1. [TRANS: *Milieu* in French means both milieu as environment and *milieu* the "middle" or "center." Canguilhem uses either meaning, by and large

referring to the meaning of milieu as environment, but at times specifically addressing the problem of an organism living (or not) in the center of its surroundings. At other times, he considers the relationship of an organism, itself a milieu (environment), to organs or tissue in it. The reader should keep the double meaning in mind, particularly when discussions of the center appear. We have largely kept the term in English, at times clarifying its meaning.]

2. Denis Diderot, *Encyclopédie de Diderot et d'Alembert: ou dictionnaire raisonné des sciences, des arts et des métiers* (Marsanne: Redon, 1999).

3. Hippolyte Taine, *Essais de critique et d'histoire* (Paris: Hachette, 1913).

4. [Sir Isaac Newton, *Opticks* (Amherst, N.Y.: Prometheus, 2003), 352, 364.—Trans.]

5. Sir Isaac Newton, *The Principia: Mathematical Principles of Natural Philosophy*, trans. I. Bernard Cohen and Anne Whitman (Berkeley: University of California Press, 1999).

6. On all these points, see Léon Bloch, *Les origines de la théorie de l'éther et la physique de Newton* (Paris: Alcan, 1908).

7. [Jean Baptiste Lamarck, "The Influence of Circumstances," in *Lamarck to Darwin: Contributions to Evolutionary Biology*, ed. Henry Lewis McKinney (Lawrence, Kan.: Coronado, 1971).—Trans.]

8. Léon Brunschvicg, *Les étapes de la philosophie mathématique* (Paris: Alcan, 1912), 508.

9. See the relation of laws to climate in Montesqieu, *De l'esprit des lois*, in *Oeuvres complètes* (Paris: Seuil, 1964), chaps. 14–18, pp. 613–640.

10. Buffon's chapter on "the degeneration of animals" in the *Histoire naturelle* (1823–1833) examines the action of the habitat and food on the animal organism. See also Buffon, *De la dégénération des animaux* (Paris: Parent Desbarres, 1868).

11. [Auguste Comte, *The Positive Philosophy of Auguste Comte*, trans. Harriet Martineau (London: George Bell & Sons, 1896), 2:9—Trans.]

12. In his behaviorist psychology, Tolman also conceives the relationships between the organism and the milieu in the form of a relation of function to variable. See André Tilquin, *Le behaviorisme* (Paris: Vrin, 1942), 439.

13. Frédéric Houssay, *Force et cause* (Paris: Flammarion, 1920); J. Constantin, "Recherches sur la Sagittaire" *Bulletin de la Société Botanique* (1885).

14. Louis Roule, *La vie des rivières* (Paris: Stock, 1948), 61.

15. We find a startling summary of this thesis in Houssay, *Force et cause*, when Houssay writes of "certain kinds of unities that we call living beings, which we designate separately, as if they really had an existence of their own, independent, whereas they have no isolated reality and they cannot be otherwise than in absolute and permanent connection with the ambient milieu, of which they are simply a localized and momentary concentration" (47).

16. René Descartes, "Fifth Discourse," in *Discourse on Method and Meditations on First Philosophy*, trans. Donald A. Cress (Indianapolis, Ind.: Hackett, 1998), 33, trans. modified.

17. This is above all the case for animals. Lamarck is more reserved concerning plants. [See, e.g., Lamarck, "The Influence of Circumstances," 13.—Trans.]

18. Charles Augustin Sainte-Beuve, *Volupté: The Sensual Man*, trans. Marilyn Gaddis Rose (Albany, N.Y.: SUNY Press, 1995), 106.

19. Charles Darwin, *On the Origin of the Species* (Cambridge, Mass.: Harvard University Press, 1964), 3. [Canguilhem has modified the text slightly—Trans.].

20. Marcel Prenant, *Darwin* (Paris: Éditions Sociales Internationales, 1938), 145–149.

21. Carl Ritter, *Comparative Geography* (Philadelphia: J. B. Lippincott & Co., 1865).

22. Alexander von Humboldt, *Kosmos* (Stuttgart: Cotta, 1845).

23. For a historical presentation of the development of this idea and a critique of its exaggerations, see Lucien Febvre, *La terre et l'evolution humaine: introduction geographique a l'histoire* (Paris: Renaissance du livre, 1922).

24. André Tilquin, *Le behaviorisme* (Paris: Vrin, 1942), 34–35. We have borrowed the bulk of the information used below from this solidly documented thesis.

25. Étienne Bonnot de Condillac, *Treatise on Sensations*, trans G. Carr (London: Favil, 1930), 3.

26. See Henri Baulig, "La géographie est-elle une science?" *Annales de Géographie* 57 (January–March 1948): 1–11; "Causalité et finalité en géomorphologie," *Geografiska Annaler* 1/2 (1949): 321–324.

27. Louis Poirier's article "L'évolution de la géographie humaine," *Critique* 8/9 (January–February 1947), provides a very interesting focus on this change of perspective in human geography.

28. On this point, see Paul Guillaume, *La psychologie de la forme* (Paris: Flammarion, 1937); and Maurice Merleau-Ponty, *La structure du comportement* (Paris: PUF, 1942).

29. Jakob von Uexküll, *Umwelt und Innenwelt der Tiere* (Berlin, 1909; 2nd ed., 1921); *Theoretische Biologie* (Berlin: Springer; 2nd ed. 1928); von Uexküll and G. Kriszat, *Streifzüge durch die Umwelten von Tieren und Menschen* (Berlin: Springer, 1934). Goldstein accepts these views of von Uexküll's only with considerable reservations: if one is unwilling to distinguish the living from its environment, all research into relations becomes in a sense impossible. Determininism disappears and is replaced by reciprocal penetration, and taking the whole into consideration kills knowledge. For knowledge to remain possible, within this organism-environment totality there must appear

a nonconventional center around which a range of relations open out. See Kurt Goldstein's "Criticism of Purely Environmental Theory: World and Environment (Milieu)," in his *The Organism* (New York: Zone, 1995), 85–90.

30. The example of the tick is taken up again, following von Uexküll, by Louis Bounoure in *L'autonomie de l'être vivant* (Paris: PUF, 1949), 143.

31. For a discussion of this thesis by Goldstein, see the conclusion of François Dagognet, *Philosophie biologique* (Paris: PUF, 1955).

32. Goldstein, *The Organism*, 388.

33. Gregor Mendel, "Versuch über Pflanzenhybriden," in *Verhandlungen des naturforschenden Vereines in Brünn*, vol. 4 (1865), treatises 3–47.

34. Albert Brachet, *La vie créatrice des formes* (Paris: Alcan, 1927), 171.

35. Maurice Caullery, *Problème de l'évolution* (Paris: Payot, 1931). One finds in Nietzsche an anticipation of these ideas. See Friedrich Nietzsche, *The Will to Power*, trans. Walter Kaufmann and R. J. Hollingdale (New York: Random House, 1967), §647, p. 345–346. In truth, the criticisms Nietzsche addresses to Darwin are more applicable to the neo-Lamarckians.

36. For a presentation of the question, see "Une discussion scientifique en U.R.S.S.," *Europe* 33/34 (1948); and also Cl. Ch. Mathon, "Quelques aspects du mitchourisme, etc." *Revue générale des sciences pures et appliquées* 3/4 (1951): 3–4. On the ideological aspect of the controversy, see Julian Huxley, *La génétique soviétique et la science mondiale* (Paris: Stock, 1950). Jean Rostand has given a good historical and critical presentation of the question in "L'offensive des mitchouriniens contre la génétique mendelienne," in his *Les grands courants de la biologie* (Paris: Gallimard, 1951), followed by a bibliography. Finally, see Raymond Hovasse, *Adaptation et évolution* (Paris: Hermann, 1951).

37. Georges L. Leclerc, Comte de Buffon, *De la dégénération des animaux* (Paris: Parent Desbarres, 1868).

38. See the article "Climate" in Diderot, *Encyclopédie de Diderot et d'Alembert*.

39. See Theodor Breiter's excellent summary of the history of Greek geography in the introduction to volume 2 (*Commentary*) of Marcus Manilius, *Astronomica* (Leipzig, 1908).

40. Blaise Pascal, *Pensées*, trans. A. J. Krailsheimer (London: Penguin, 1966), 88–95.

41. [Ibid., 90.—Trans.]

42. [Ibid., 92–93.—Trans.]

43. [Ibid., 93.—Trans.]

44. [Ibid., 89—Trans.]

45. Dietrich Mahnke, *Unendliche Sphäre und Altmittelpunkt* (Halle: Niemeyer, 1937); this author dedicates several very interesting pages to the use and signification of this expression in Leibniz and Pascal. According to Eugène Havet, Pascal borrowed the expression either from Melle de Gournay (the

preface to the 1595 edition of Montaigne's *Essais*) or from Rabelais's *Third Book of Pantagruel* (chap. 13).

46. See Alexandre Koyré, *La philosophie de Jacob Boehme* (Paris: Vrin, 1929), 378–379, 504; also his "The Significance of the Newtonian Synthesis," *Archives internationales d'histoire des sciences* 11 (1950), reprinted in Koyré, *Newtonian Studies* (Chicago: University of Chicago Press, 1965), 3–25. [See also Koyré, *From the Closed World to the Infinite Universe* (Baltimore, Md.: Johns Hopkins University Press, 1957), chap. 9—Trans.].

47. Edouard Claparède, preface to Frederik Jacobus Johannes Buytendijk, *Psychologie des animaux* (Paris: Payot, 1928).

THE HERMENEUTICS OF THE SUBJECT
MICHEL FOUCAULT

1. From 1982, Foucault, who previously had both lectured and held a seminar, decided to give up the seminar and just lecture, but for two hours.

2. See the summary of the 1980–1981 course at the Collège de France in *Dits et écrits, 1954–1988*, eds. D. Defert and F. Ewald (Paris: Gallimard, 1994), 4:213–218; English translation by Robert Hurley, "Subjectivity and Truth," in *The Essential Works of Michel Foucault, 1954–1984*, vol. 1: *Ethics: Subjectivity and Truth*, ed. Paul Rabinow (New York: The New Press, 1997), 87–92.

3. For the first elaboration of this theme, see the lecture of January 28, 1981, but more especially Michel Foucault, *L'usage des plaisirs* (Paris: Gallimard, 1984), 38–52. By *aphrodisia* Foucault understands an *experience*, which is a *historical* experience: the Greek experience of pleasures as opposed to the Christian experience of the *flesh* and the modern experience of *sexuality*. The *aphrodisia* are identified as the "ethical substance" of ancient morality.

4. In the first lecture of the 1981 course ("Subjectivité et verité," January 7, 1981) Foucault states that what is at stake in his research is whether it was not precisely paganism that developed the strictness and sense of decency of our moral code (which, furthermore, would problematize the break between Christianity and paganism in the field of a history of morality).

5. In the 1981 lectures there are no analyses explicitly concerned with the care of the self, but there are lengthy analyses dealing with the arts of existence and processes of subjectivation (the lectures of January 13, March 25, and April 1). However, generally speaking, the 1981 course continues to focus exclusively on the status of the aphrodisiac in pagan ethics of the first two centuries AD while maintaining that we cannot speak of subjectivity in the Greek world, the ethical elements being determined as *bios* (mode of life).

6. All the important texts of Cicero, Lucretius, and Seneca on these problems of translation have been brought together by Carlos Levy in his article "Du grec au latin," in *Le discours philosophique* (Paris: PUF, 1998), 1145–1154.

7. "If I do everything in my own interest, it is because the interest I have in myself comes before everything else (*Si omnia propter curam mei facio ante omnia est mei cura*)." Seneca *Letters* CXXI.17.

8. See P. Courcelle, *Connais-toi-même, de Socrate à saint Bernard* (Paris: Etudes augustiniennes, 1974), 3 vols.

9. Epictetus *Discourses* III.i.18–19.

10. For the Greeks, Delphi was the geographical center of the world (*omphalos*; the world's navel), where the two eagles sent by Zeus from the opposite sides of the Earth's circumference came together. Delphi became an important religious center at the end of the eighth century BC (the sanctuary of Apollo from which Python delivered oracles) and continued to be so until the end of the fourth century AD, extending its audience to the entire Roman world.

11. W. H. Roscher, "Weiteres über die Bedeutung des E[ggua] zu Delphi und die übrigen *grammata Delphika*," *Philologus* 60 (1901): 81–101.

12. The second maxim is: *eggua, para d'atē*. See Plutarch's statement in *Dinner of the Seven Wise Men* 164b: "Until I have learned it from these gentlemen, I won't be able to explain to you the meaning of the precepts *Not too much* and *Know yourself*, and the famous maxim which has stopped so many from getting married, has made so many others mistrustful and others silent: Commitment brings misfortune (*eggua para d'ata*)."

13. J. Defradas, *Les thèmes de la propragande delphique* (Paris: Klincksieck, 1954), chap. 3, "La sagesse delphique," 268–283.

14. "Then Socrates demanded: 'Tell me, Euthydemus, have you ever been to Delphi?'

'Yes, by Zeus,' Euthydemus answered, 'I have been twice.'

'Then did you notice somewhere on the temple the inscription: Know yourself?'

'Yes.'

'Did you just idly glance at it, or did you pay attention to it and try to examine who you are?'"

Xenophon *Memorabilia* IV.II.24.

15. For his lectures Foucault usually uses the Belles Lettres edition (otherwise called the Budé edition) that enables him to have the original Greek or Latin facing the translation. This is why for the important terms and passages he accompanies his reading with references to the text in the original language. Moreover, when Foucault reads French translations in this way, he does not always follow them to the letter, but adapts them to the demands of oral style, multiplying logical connectors ("and," "or," "that is to say," "well," etc.) or giving reminders of the preceding arguments. Usually we restore the original French translation while indicating, in the text, significant additions (followed by "M.F.") in brackets.

16. Plato *Apology* 29d.

17. Foucault here cuts a sentence from 30a: "If it seems clear that, despite what he says, he does not possess virtue, I shall reproach him for attaching less value to what has the most value and more value to what has the least." Ibid.

18. Ibid., 30a.

19. "I tell you, being what I am, it is not to me that you do the most wrong if you condemn me to death, but to yourselves." Ibid., 30c.

20. Foucault refers here to a development of the exposition from 31a to 31c.

21. In 35–37a, on being told of his condemnation to death, Socrates proposes an alternative penalty. Actually, in the kind of trial Socrates undergoes, there is no penalty fixed by law: it is up to the judges to determine the penalty. The penalty demanded by the accusers (and indicated in the charge) was death, and the judges acknowledge that Socrates is guilty of the misdeeds of which he is accused and therefore liable to incur this penalty. However, at this moment of the trial, Socrates, recognized as guilty, must propose an alternative penalty. It is only after this that the judges must fix a punishment for the accused on the basis of the penal proposals of the two parties. For further details see C. Mossé, *Le procès de Socrate* (Brussels: Éd. Complexe, 1996), as well as the lengthy introduction by L. Brisson to his edition of the *Apologie de Socrate* (Paris: Garnier Flammarion, 1997).

22. Plato *Apology* 36b–d.

23. This alludes to the famous passage of 28d: "The principle, Athenians, is this. Someone who occupies a post (*taxē*), whether chosen by himself as most honorable or placed there by a commander, has to my mind the duty to remain in place whatever the risk, without thought of death or danger, rather than sacrifice honor." Epictetus praises steadfastness in one's post as the philosophical attitude par excellence. See, for example, *Discourses* I.xi.24 and III.xxiv.36 and 95, in which Epictetus alternates between the terms *taxis* and *khōra*. See also the end of Seneca's *On the Firmness of the Wise Man* XIX.4: "Defend the post (locum) that nature has assigned you. You ask what post? That of a man."

24. Socrates warns the Athenians of what will happen if they condemn him to death: "You will spend the rest of your life asleep." Plato *Apology* 31a.

25. "If you put me to death you will not easily find another man . . . attached to you by the will of the gods in order to stimulate you like a horsefly stimulates a horse." Plato *Apology* 30e.

26. "Did Socrates manage to persuade all those who came to him to take care of themselves (*epimeleisthai heatōn*)?" Epictetus *Discourses* III.i.19.

27. It is found in the *Letter to Menoeceus* 122. More exactly the text says: "For no-one is it ever too early or too late for ensuring the soul's health . . . So young and old should practice philosophy." This quotation is taken up by Foucault in Michel Foucault, *Histoire de la sexualité*, vol. 3, *Le souci de soi* (Paris:

Gallimard, 1984), 60; English translation by Robert Hurley, *The Care of the Self* (New York: Pantheon, 1985), 46.

28. Actually, the Greek text has "*to kata psukhēn hugiainon.*" The verb *therapeuein* appears only once in Epicurus, in *Vatican Saying* 55: "We should treat (*therapeuteon*) misfortunes with the grateful memory of what we have lost and with the knowledge that what has come about cannot be undone."

29. The center of gravity for the whole of this theme is Epicurus's phrase: "The discourse of the philosopher who does not treat any human affection is empty. Just as a doctor who does not get rid of bodily diseases is useless, so also is a philosophy if it does not get rid of the affection of the soul (221 Us.)." Translated by A.-J. Voelke in his *La philosophie comme thérapie de l'âme* (Paris: Éd. du Cerf, 1993), 36. In the same work, see the articles: "Santé de l'âme et bonheur de la raison. La fonction thérapeutique de la philosophie dans épicurisme" and "Opinions vides et troubles de l'âme: la médication épicurienne."

30. Seneca *On Benefits* VII.i.3–7. This text is analyzed at length in the lecture of February 10, second hour.

31. For a conceptualization of the notion of the culture of the self, see the lecture of January 6, first hour.

32. On the concept of the event in Foucault, see "Nietzsche, la généalogie, l'histoire" (1971) in *Dits et écrits*, 2:136, for the Nietzschean roots of the concept; and "Mon corps, ce papier, ce feu" in *Dits et écrits*, 2:260, on the polemical value of the event in thought against a Derridean metaphysics of the originary (English translations by Robert Hurley and others, as "Nietzsche, Genealogy, History" and "My Body, This Paper, This Fire," in *Essential Works of Foucault, 1954–1984*, vol. 2: *Aesthetics, Method, and Epistemology*, ed., J. D. Faubion [New York: New Press, 2000]), "Tale ronde du 20 mai 1978" for the program of an *événementialisation* of historical knowledge, *Dits et écrits*, 4:23; and in particular, "Polemique, politique, et problematisations" in *Dits et écrits*, vol. 4, concerning the distinctiveness of the history of thought (translated by Lydia Davis as "Polemics, Politics, and Problematizations: An Interview with Michel Foucault" in *Ethics: Subjectivity and Truth*).

33. "Considering the seventh day to be very holy and a great festival, they accord it a special honor: on this day, after caring for the soul (*tēs psukhēs epimeleian*), they anoint their bodies with oil." Philo of Alexandria *On the Contemplative Life* 477M, IV.36.

34. "Then we will contemplate the same objects as [the soul of the universe], because we also will be well prepared thanks to our nature and our efforts (*epimeleias*)." Plotinus *Enneads* II.9.18.

35. "The law eliminates fate by teaching that virtue is taught and develops if one applies oneself to it (*ex epimeleias prosginomenēn*)." Methodius of Olympus *The Banquet* 172c.

36. "*Hote toinun hē agan hautē tou sōmatos epimeleia autō te alusitelēs to sōmati, kai pros tēn psukhēn empodion esti; to ge hupopeptōkenai toutō kai therapeuein mania*

saphēs" ("When excessive care for the body becomes useless for the body and harmful to the soul, submitting to it and attaching oneself to it seems an obvious madness"). Basil of Caesarea *Sermo de legendis libris gentilium* 584d, in J.-P. Migne, ed. *Patrologie grecque* (SEU Petit-Montrouge, 1857), vol. 31.

37. "Now that [Moses] had raised himself to the highest level in the virtues of the soul, both by lengthy application (*makras epimeleias*) and by knowledge from on high, it is, rather, a happy and peaceful encounter that he has with his brother . . . The help given by God to our nature . . . only appears . . . when we are sufficiently familiarized with the life from on high through progress and application (*epimeleias*)." Grégoire de Nysse [Gregory of Nyssa], *La vie de Moïse, ou traité de la perfection en matière de la vertue*, trans. J. Daniélou (Paris: Éd. du Cerf, 1965), 337c–d, 43–44, pp. 130–131; see also 55 in 341b, setting out the requirement of a "long and serious study (*toiautēs kai tosautēs epimeleias*)," 138.

38. "But now I have returned here to this same grace, joined by love to my master; also strengthen in me what is ordered and stable in this grace, you the friends of my fiancé, who, by your cares (epimeleias) and attention, preserve the impulse in me towards the divine." Grégoire de Nysse, *Le cantique des cantiques*, trans. C. Bouchet (Paris: Migne, 1990), 106.

39. "*Ei oun apokluseias palin di'epimeleias biou ton epiplasthenta tē kardia sou rupon, analampsei soi to theoeidēs kallos* (If, on the other hand, you purify the dregs spread out in your heart by taking care of your life, the divine beauty will shine within you)." Gregory of Nyssa *De Beatitudinibus* Oratio VI, in *Patrologie grecque*, 44:1272a.

40. Gregory of Nyssa *Treatise on Virginity*. See in the same book the parable of the lost drachma (300c–301c, XII), often cited by Foucault to illustrate the care of the self. See the lecture "Technologies of the Self," in *Ethics: Subjectivity and Truth*, 227); "Les techniques de soi" in *Dits et écrits*, 4:787: "By filth, we should understand, I think, the taint of the flesh: when one has 'swept it away' and cleared it by the 'care' (*epimeleia*) that one takes of one's life, the object appears in broad daylight." 301 XII, 3.

41. In an interview in January 1984 Foucault notes that in this text by Gregory of Nyssa (303c–305c, XIII) the care of the self is essentially defined as "the renunciation of all earthly attachments. It is the renunciation of all that may be love of self, of attachment to an earthly self." "L'éthique du souci de soi comme pratique de la liberté," in *Dits et écrits*, 4:716; English translation by P. Aranov and D. McGrawth, "The Ethics of the Concern for Self as a Practice of Freedom," in *Ethics: Subjectivity and Truth*, 288.

42. On the meaning of *meletē*, see the lecture of March 3, second hour; and March 17, first hour.

43. On the techniques of meditation, and the meditation on death in particular, see the lectures of February 27, second hour; March 3, first hour; and March 24, second hour.

44. On examination of conscience see the lecture of March 24, second hour.

45. On the technique of screening representations, in Marcus Aurelius in particular, and in comparison with the examination of ideas in Cassian, see the lecture of February 24, first hour.

46. In "moral dandyism" we can see a reference to Baudelaire (see Foucault's pages on "the attitude of modernity" and the Baudelairean ethos in "What Is Enlightenment?" in *Ethics: Subjectivity and Truth*, 310–312 [French version: "Qu'est-ce que les Lumières?" in *Dits et écrits*, 4:568–571]) and in the "aesthetic stage" there is a clear allusion to Kierkegaard's existential triptych (aesthetic, ethical, and religious stages), the aesthetic sphere (embodied by the Wandering Jew, Faust, and Don Juan) being that of the individual who exhausts the moments of an indefinite quest as so many fragile atoms of pleasure (it is irony that allows transition to the ethical). Foucault was a great reader of Kierkegaard, although he hardly ever mentions this author, who nonetheless had for him an importance as secret as it was decisive.

47. This thesis of the Hellenistic and Roman philosopher no longer finding the basis for the free use of his moral and political action in the new sociopolitical conditions (as if the Greek city-state had always been its natural element), and finding in the self a last resort into which to withdraw, became a *topos*, if not unchallenged self-evidence of the history of philosophy (shared by Bréhier, Festugière, and others). During the second half of the century, the articles on epigraphy and the teaching of a famous scientist with an international audience, Louis Robert (*"Opera minora selecta." Épigraphie et antiquité grecques* [Amsterdam: Hakkert, 1989], 6:715), made this vision of the Greek lost in a world which was too big and in which he was deprived of his city state outmoded (I owe all this information to Paul Veyne). This thesis of the obliteration of the city-state in the Hellenistic period is thus strongly challenged by, among others, Foucault in *Le souci de soi* (*The Care of the Self*, part 3, chap. 2, "The Political Game," 81–95; and see also 41–43). For Foucault it is primarily a question of challenging the thesis of a breakup of the political framework of the city-state in the Hellenistic monarchies (81–83) and then of showing (and again in this course) that the care of the self is basically defined as a mode of living rather than as an individualistic resort ("The care of the self . . . appears then as an intensification of social relations," 53). P. Hadot, in *Qu'est-ce que la philosophie antique?* (Paris: Gallimard, 1995), 146–147, traces this prejudice of a disappearance of the Greek city-state back to a work by G. Murray, *Four Stages of Greek Religion* (New York: Columbia University Press, 1912).

48. Descartes, *Méditations sur la philosophie première* (1641), in Œuvres (Paris: Gallimard/Bibliotheque de la Pleiade, 1952); English translation by John Cottingham, in Descartes, *Meditations on First Philosophy*, ed. John Cottingham (Cambridge: Cambridge University Press, 1996).

49. Gnosticism represents an esoteric philosophico-religious movement that developed in the first centuries AD. This extremely widespread move-

ment, which is difficult to delimit and define, was rejected both by the Church Fathers and by philosophy inspired by Platonism. The *"gnosis"* (from the Greek *gnōsis*: knowledge) designates an esoteric knowledge that offers salvation to whomever has access to it, and for the initiated it represents knowledge of his origin and destination as well as the secrets and mysteries of the higher world (bringing the promise of a heavenly voyage), uncovered on the basis of secret exegetical traditions. In this sense of a salvationist, initiatory, and symbolic knowledge, the "gnosis" covers a vast set of Judeo-Christian speculations based on the Bible. The "Gnostic" movement, through the revelation of a supernatural knowledge, thus promises liberation of the soul and victory over the evil cosmic power. For a literary reference see Michel Foucault, "La prose d'Actéon" in *Dits et écrits*, 1:326. It is likely, as A. I. Davidson has suggested to me, that Foucault was familiar with the studies of H. C. Puech on this subject (See *Sur le manichéisme et autres essais* [Paris: Flammarion, 1979]).

50. "The" philosopher is how Aquinas designates Aristotle in his commentaries.

51. In the classification of the conditions of knowledge that follow we find, like a muffled echo, what Foucault called "procedures of limitation of discourse" in his inaugural lecture at the College de France, *L'ordre du discours* (Paris, Gallimard, 1971). However, in 1970 the fundamental element was discourse, as an anonymous and blank sheet, whereas everything here is structured around the articulation of the "subject" and "truth."

52. We can recognize here an echo of the famous analysis devoted to the *Meditations* in Foucault's *Histoire de la folie* (Paris: Gallimard/Tel, 1972). In the exercise of doubt Descartes encounters the vertigo of madness as a reason for doubting, and he excludes it a priori, refuses to countenance it, preferring the gentle ambiguities of the dream: "madness is excluded by the doubting subject" (7). Derrida immediately challenges this thesis in "Cogito et Histoire de la folie" (in *L'écriture et la différence* [Paris: Éd. du Seuil, 1967]; English translation by A. Bass, "Cogito and the History of Madness," in *Writing and Difference* [London: Routledge and Kegan Paul, 1978]), which takes up a lecture delivered on March 4, 1963, at the Collège Philosophique, showing that the peculiarity of the Cartesian Cogito is precisely to take on the risk of a "total madness" by resorting to the hypothesis of the evil genius (81–82; English translation, 52–53). We know that Foucault, openly stung by this criticism, some years later published a masterly response, raising a specialist quarrel to the level of an ontological debate through a rigorous textual explanation ("My Body, This Paper, This Fire" and "Réponse à Derrida," in *Dits et écrits*, vol. 2). Thus was born what is called the "Foucault/Derrida polemic" about Descartes' *Meditations*.

53. Foucault examines the Faust myth at greater length in the lecture of February 24, second hour.

54. B. Spinoza, "Tractatus de intellectus emendatione," in *Benedicti de Spinoza Opera quotquot reperta sunt*, ed. J. Van Vloten and J. P. N Land (The

Hague, 1882–1884); English translation by R. H. M. Elwes, "On the Improvement of the Understanding," in *Works of Spinoza*, vol. 2 (New York: Dover, 1955).

55. E. Husserl, *Die Krisis der europäischen Wissenschaften und die transzendentale Phänomenologie* (Belgrade: Philosophia, 1936); English translation by D. Carr, *The Crisis of European Sciences and Transcendental Phenomenology* (Evanston, Ill.: Northwestern University Press, 1970).

56. In this period Foucault identified himself as an heir to this tradition that he recognized as that of "modern" philosophy. See Michel Foucault, "Qu'est-ce que les Lumières?" in *Dits et écrits*, 4:687–688; English translation by Colin Gordon, "Kant on Enlightenment and Revolution," in *Economy and Society*, vol. 15, no. 1 (London: Routledge and Kegan Paul, 1986), 403–404; and "The Political Technology of Individuals," in *The Essential Works of Foucault, 1954–1984*, vol. 3: *Power*, ed. J. D. Faubion (New York: New Press, 2000), 403–404; French translation by P.-E. Dauzat, "La technologie politique des individus," in *Dits et écrits*, 4:813–814.

57. G. W. F. Hegel, *Phänomenologie des Geistes* (Wurtzbourg: Anton Goebhardt, 1807); French translation by J. Hyppolite, *Phénomenologie de l'esprit* (Paris: Aubier-Montaigne, 1941); English translation by A. V. Miller, *The Phenomenology of Spirit* (Oxford: Clarendon, 1979).

58. On Lacan's reopening of the question of the subject, see *Dits et écrits*, 3:590, 4:204–205, and 4:435. For Lacan's texts going in this direction, see "Fonction et champ de la parole et du langage en psychanalyse" (1953), "Subversion du sujet et dialectique du désir dans l'inconscient freudien" (1960), "La science et la vérité" (1965), and "Du sujet enfin la question" (1966), all in Jacques Lacan, *Écrits* (Paris: Le Seuil, 1966) (English translations by Alan Sheridan in *Ecrits: A Selection* [London: Tavistock/Routledge, 1989]); *Le séminaire I: les écrits techniques de Freud (1953–1954)* (Paris: Le Seuil, 1975), 287–299; *Le séminaire II: le moi dans la théorie de Freud et dans la technique de la psychanalyse (1954–1955)* (Paris: Le Seuil, 1978); *Les quatre concepts fondamentaux de la psychanalyse (1964)* (Paris: Le Seuil, 1973), 31–41, 125–135 (English translation by Alan Sheridan, *The Four Fundamental Concepts of Psychoanalysis* [London: Hogarth, 1977], 29–41, 136–149); "Réponse à des étudiants en philosophie sur l'objet de la psychanalyse," *Cahiers pour l'analyse* 3 (1966): 5–13; "La méprise du sujet supposé savoir," *Scilicet* 1 (Paris: Le Seuil, 1968); *Le séminaire XX: encore* (Paris: Le Seuil, 1975), 83–91; "Le symptôme," *Scilicet* 6/7 (Paris: Le Seuil, 1976), 42–52. I am indebted to J. Lagrange and to M. Benani for this note.

59. This third moment will not be developed in this year's course, or in the following year.

60. "As someone asked why they entrusted work in the fields to the helots, instead of taking care of them themselves (*kai ouk autoi epimelountai*). 'Because,'

he answered, 'it was not in order to take care of them that we acquired them, but to take care of ourselves (*ou toutōn epimelomenoi all'hautōn*).'" Plutarch *Sayings of Spartans* 217a. See the summary of this example in *Le souci de soi*, 58; *The Care of the Self*, 44.

61. They are examined in the second hour of the lecture of January 13.

62. All of this takes place in the beginning of the text, *Alcibiades* 1, from 103a to 105e.

63. Foucault is thinking here of Achilles' double destiny: "For my mother Thetis the goddess of the silver feet tells me/I carry two sorts of destiny towards the day of my death. Either,/if I stay here and fight beside the city of the Trojans,/my return home is gone, but my glory shall be everlasting;/but if I return home to the beloved land of my fathers,/the excellence of glory is gone, but there will be a long life/left for me, and my end in death will not come to me quickly." Homer, *The Iliad*, translation by Richmond Lattimore (Chicago: 1961), book IX, 410–416, p. 209. French translation by P. Mazon, *Iliade* (Paris: Les Belles Lettres, 1937), 67.

64. Plato *Alcibiades* 104a.

65. Through his father Clinias, Alcibiades was a member of the *genos* of the "Eupatrids" (i.e., "those of good fathers"), a family of aristocrats and big landowners who dominated Athens politically from the archaic period. His mother (daughter of Megacles, a victim of ostracism) belongs to the family of Alcmaeonids, who undoubtedly played the most decisive role in the political history of classical Athens.

66. Plato *Alcibiades* 104b.

67. The problem of the critical age of boys was broached by Foucault in the lecture of January 28, 1981, in particular, which was devoted to the structuring of the ethical perception of *aphrodisia* (principle of socio-sexual isomorphism and principle of activity) and the problem raised within this framework by the love of young boys from good families.

68. Xenophon *Memorabilia* III.vii.

69. More precisely, the Greek text has *alla diateinou mallon pros to seautō prosekhein*.

70. This passage is found in Plato *Alcibiades* 119a–124b.

71. "Ah, naïve child, believe me and the words inscribed at Delphi: 'Know yourself.'" Plato *Alcibiades* 124b.

72. Ibid., 125d.

73. Ibid., 126c.

74. Ibid., 127d.

75. Ibid., 127e.

76. Plato *Apology* 30a.

77. According to Diogenes Laertius, *Lives of Eminent Philosophers* III.57–62, the catalogue of Thrasylus (astrologer of Tiberius and philosopher at Nero's

court in the first century AD) adopts the division of Plato's dialogues into tetralogies and for each dialogue fixes a title, which usually corresponds to the name of Socrates' principal interlocutor—but it may be that this way of designating the dialogues goes back to Plato himself—and a second title indicating the main subject matter.

78. The expression is found in Plato *Alcibiades* 129b.

THE COURAGE OF THE TRUTH

MICHEL FOUCAULT

1. Foucault's lectures at the Collège de France became devoted to ancient thought from January 1981, with "Subjectivité et Vérité." See the course summary in *Dits et écrits, 1954–1988*, ed. D. Defert and F. Ewald (Paris: Gallimard, 1994), 4: 213–218; English translation by Robert Hurley, "Subjectivity and Truth," in *The Essential Works of Michel Foucault, 1954–1984*, vol. 1: *Ethics: Subjectivity and Truth*, ed. Paul Rabinow (New York: The New Press, 1997), 87–92.

2. Already in the previous year, faced with the same difficulty, Foucault had called for the formation, in parallel with the main course, of a small working group which would be made up exclusively of researchers working on closely related themes. See *Le gouvernement de soi et des autres. Cours au Collège de France, 1982–1983*, ed. Frédéric Gros (Paris: Gallimard-Le Seuil, 2008), 3 (lecture of January 5, 1983, first hour) and 68 (lecture of January 12, 1983, second hour); English translation by Graham Burchell, *The Government of Self and Others: Lectures at the Collège de France, 1982–1983*, ed. Frédéric Gros (London: Palgrave Macmillan, 2010), 1, 70.

3. On the concept of alethurgy, see the Collège de France lectures of January 23 and 30, 1980: "by creating the fictional word *alēthourgia* from *alēthourgēs*, we could call 'alethurgy' (manifestation of truth) the set of possible procedures, verbal or otherwise, by which one brings to light what is posited as true, as opposed to the false, the hidden, the unspeakable, the unforeseeable, or the forgotten. We could call 'alethurgy' that set of procedures and say that there is no exercise of power without something like an alethurgy" (lecture of January 23).

4. M. Foucault, *Histoire de la folie à l'âge classique* (Paris: Plon, 1961; Gallimard, 1972); English translation by Jonathan Murphy and Jean Khalfa as *History of Madness* (London: Routledge, 2005); M. Foucault, *Surveiller et punir* (Paris: Gallimard, 1975); English translation by Alan Sheridan as *Discipline and Punish: Birth of the Prison* (New York: Pantheon, 1977).

5. M. Foucault, *Les mots et les choses* (Paris: Gallimard, 1966); English translation by Alan Sheridan as *The Order of Things* (New York: Pantheon, 1970). For a similar presentation of the analysis of the "speaking, laboring, living subject,"

see the notice "Foucault" in *Dits et écrits*, 4:633; English translation by Robert Hurley as "Foucault" in *The Essential Works of Foucault, 1954–1984*, vol. 2: *Aesthetics, Method, and Epistemology*, ed. James D. Faubion (New York: The New Press, 1998), 460.

6. The same systematic presentation of his oeuvre in the form of a triptych is found in the first lecture of 1983, *Le gouvernement de soi et des autres*, 4–7; *The Government of Self and Others*, 2–5.

7. On the examination of conscience as spiritual exercise, see the lecture of March 24, 1982, second hour, in M. Foucault, *L'herméneutique du sujet. Cours au Collège de France, 1981–1982*, ed. Frédéric Gros (Paris: Gallimard-Le Seuil, 2001), 460–464, as well as the lectures of January 20 and 27, 1982, ibid., 86–87, 146–149, and 151–157; English translation by Graham Burchell as *The Hermeneutics of the Subject: Lectures at the Collège de France, 1981–1982* (New York: Palgrave Macmillan, 2001), 480–485, 89–90, 151–154, and 157–163.

8. On correspondence as spiritual exercise, see the lectures of January 20 and 27, 1982, ibid., Fr. 86–87, 146–149, and 151–157; Eng. 89–90, 151–154, and 157–163.

9. On the *hupomnēmata* and other writing exercises, see the lecture of March 3, 1982, in ibid., Fr. 341–345; Eng. 358–362, as well as "L'écriture de soi" in *Dits et écrits*, 4:415–430; English translation by Robert Hurley as "Self Writing" in *The Essential Works of Foucault*, 1:207–222.

10. See *L'herméneutique du sujet*; *The Hermeneutics of the Subject*, all the lectures of January 1982.

11. On this concept, see the lecture of February 3, 1982, ibid., Fr. 172–173; Eng. 179–180. One could also consult the article by Pierre Hadot in *Michel Foucault philosophe* [Rencontre internationale Paris, January 9–11, 1988], ed. l'Association pour le Centre Michel Foucault (Paris: Le Seuil, 1989); English translation by Timothy J. Armstrong in *Michel Foucault Philosopher* (Hempstead: Harvester Wheatsheaf, 1992).

12. On this history see the lecture of February 19, 1975, in M. Foucault, *Les anormaux. Cours au Collège de France, 1974–1975*, ed. Valerio Marchetti and Antonella Salomoni (Paris: Gallimard-Le Seuil, 1999), 161–171; English translation by Graham Burchell as *Abnormal: Lectures at the Collège de France, 1974–1975* (New York: Picador, 2003), 174–184. See also M. Foucault, *La volonté de savoir* (Paris: Gallimard, 1976), esp. the chapter "L'incitation aux discours"; English translation by Robert Hurley as *The History of Sexuality*, vol. 1: *An Introduction* (New York: Pantheon, 1978), "The Incitement to Discourse."

13. On this concept see the lecture of February 22, 1978, in M. Foucault, *Sécurité, territoire, population: Cours au Collège de France, 1977–1978*, ed. Michel Senellart (Paris: Gallimard-Le Seuil, 2004); English translation by Graham Burchell as *Security, Territory, Population: Lectures at the Collège de France, 1977–1978* (Basingstoke: Palgrave Macmillan, 2007). See also "Omnes et

singulatim," trans. P. E. Dauzat, in *Dits et écrits*, 4:136–147; original English version, "'Omnes et singulatim': Toward a Critique of Political Reason," in *The Essential Works of Foucault, 1954–1984*, vol. 3: *Power*, ed. James D. Faubion (New York: New Press, 2000), 300–315.

14. Galen, *Traité des passions de l'âme et des erreurs*, trans. R. Van der Elst (Paris: Delagrave, 1914); English translation by P. W. Harkins as Galen, *On the Passions and Errors of the Soul* (Columbus: Ohio State University Press, 1963). For Foucault's analysis of this text, see *L'herméneutique du sujet*, 378–382; *The Hermeneutics of the Subject*, 395–398; and lecture of January 12, 1983, first hour, in *Le gouvernement de soi et des autres*, 43–45; *The Government of Self and Others*, 43–45.

15. On the organization of Epictetus's school, see the lecture of January 27, 1982, first hour, in *L'herméneutique du sujet*, 133–137; *The Hermeneutics of the Subject*, 138–142.

16. On this figure see, ibid., Fr. 137–138; Eng. 142–144.

17. On this "medical" dimension of the care of the soul, see Foucault's clarifications in the lecture of January 20, 1982, first hour, ibid., Fr. 93–96; Eng. 97–100.

18. See the chapter "Du régime en général," in M. Foucault, *L'usage des plaisirs* (Paris: Gallimard, 1984); English translation by Robert Hurley, "Regimen in General," in *The Use of Pleasure* (New York: Random House, 1985).

19. On *libertas* (the Latin translation of *parrhēsia*) in Seneca, see the lecture of March 10, 1982, second hour, in *L'herméneutique du sujet*, 382–388; *The Hermeneutics of the Subject*, 398–405.

20. For a first analysis of the treatise by Philodemus, see ibid., Fr. 370–374; Eng. 387–391, and the lecture of January 12, 1983, first hour, *Le gouvernement de soi et des autres*, 45–46; *The Government of Self and Others*, 41–59.

21. See on this text, *L'herméneutique du sujet*, 357–358; *The Hermeneutics of the Subject*, 373–374, and the lecture of March 2, 1983, in *Le gouvernment de soi et des autres*; *The Government of Self and Others*.

22. See above, note 14.

23. For the history of this "divergence," see the lecture of March 2, 1983, first hour, in *Le gouvernement de soi et des autres*, 277–282; *The Government of Self and Others*, 299–323.

24. *Sécurité, territoire, population*; *Security, Territory, Population*.

25. For a similar presentation of *parrhēsia* as the node of the three major axes of research, see *Le gouvernement de soi et des autres*, 42; *The Government of Self and Others*, 42.

26. For a "long version" of the presentation of his method, see the start of the lecture of January 5, 1983, ibid., Fr. 3–8; Eng. 1–6.

27. See the first definitions in March 1982, *L'herméneutique du sujet*, 348; *The Hermeneutics of the Subject*, 366; and January 1983, *Le gouvernement de soi et des autres*, 42–43; *The Government of Self and Others*, 42–43.

28. Démosthène, *Plaidoyers politiques*, trans. G. Mathieu (Paris: Les Belles Lettres, 1972), 3:76, §237: "It is necessary, Athenians, to speak with frankness (*meta parrhēsias*) without holding back before anything"; English translation by Douglas M. MacDowell, in Demosthenes, *On the False Embassy* (Oration 19), ed. Douglas M. MacDowell (Oxford: Oxford University Press, 2000), 157: "It's necessary to speak to them freely, men of Athens, without restraint."

29. Démosthène, *Première philippique*, §50, in *Harangues*, trans. M. Croiset (Paris: Les Belles Lettres, 1965), 1:49: "I have just stated my thought to you without concealing anything (*panth'haplōs ouden huposteilamenos, peparrhēsiasmai*)"; English translation by J. H. Vince, Demosthenes, *First Philippic*, in *Demosthenes, Orations* (Cambridge, Mass.: Harvard University Press, Loeb Classical Library, 1930), 1:99, §51: "to-day, keeping nothing back, I have given free utterance to my plain sentiments."

30. Isocrate, *Busiris*, §40, in *Discours*, trans. G. Mathieu and E. Brémond (Paris: Les Belles Lettres, 1972), 1:198: "We are not to say everything concerning the gods (*tēs d'eis tous theous parrhēsias oligōrēsomen*)"; English translation by Larue Van Hook, Isocrates *Busiris*, in Isocrates, Loeb Classical Library (Cambridge, Mass.: Harvard University Press, 1945) 3:125: "we [shall] disregard loose-tongued vilification of the gods."

31. For a first analysis of this passage of *The Republic*, book VIII, 557a–b et seq., see the lecture of February 9, 1983, first hour, in *Le gouvernement de soi et des autres*, 181–185; *The Government of Self and Others*, 197–201.

32. Démosthène, *Seconde Philippique*, in *Harangues*, trans. M. Croiset (Paris: Les Belles Lettres, 1965), 2:34, §32; English translation by J. H. Vince, Demosthenes, *The Second Phillipic*, in Demosthenes, *Orations*, 1:141: "It is not that by descending to abuse I may lay myself open to retaliation in your presence."

33. Ibid., §31, Fr. 34: "Ah! I will speak to you with open heart, I call on the gods to witness it, I wish to conceal nothing (*egō nē tous theous talēthē meta parrhēsias erō pros humas kai ouk apokrupsōmai*)"; Eng. 141: "I vow that I shall boldly tell you the whole truth and keep nothing back."

34. *Première Philippique*, §51, p. 49: "In fact, I do not know what the consequences of my proposals will be for me"; *First Philippic*, p. 99: "in the uncertainty of what the result of my proposals may be for myself."

35. On this story and Foucault's analysis of it in terms of *parrhēsia*, see the lecture of January 12, 1983, first hour, *Le gouvernement de soi et des autres*, 47–52; *The Government of Self and Others*, 47–52.

36. Aristotle, *Nicomachean Ethics*, trans. W. D. Ross, rev. J. O. Urmson, book IV, 1124b, 26–29, in *The Complete Works of Aristotle: The Revised Oxford Translation*, ed. Jonathan Barnes (Princeton, N.J.: Princeton University Press, 1984), 2:1775: "He must also be open in his hate and in his love (for to conceal one's feelings is a mark of timidity), and he must care more for truth than for what people will think, and must speak and act openly; for he is free of speech

because he is contemptuous, and he is given to telling the truth, except when he speaks in irony to the vulgar."

37. "Heraclitus," in Diogène Laërce, *Vie, doctrines et sentences des philosophes illustres*, ed. and trans. R. Genaille (Paris: Garnier-Flammarion, 1965), 2:163; and Diogène Laërce, *Vie et doctrines des philosophes illustres*, ed. and trans. M.-O. Goulet-Cazé (Paris: Le Livre de poche, 1999), book IX, §2, p. 1048; English translation by R. D. Hicks, "Heraclitus," in Diogenes Laertius, *Lives of Eminent Philosophers*, book IX, Loeb Classical Library (Cambridge, Mass.: Harvard University Press, 1925), 2:411: "We will have none who is worthiest among us."

38. Ibid.; Eng.: "'Why, you rascals,' he said, 'are you astonished? Is it not better to do this than to take part in your civil life?'"

39. Ibid.

40. Ibid., Fr. 165 (Goulet-Cazé trans. IX, 12, p. 1050); Eng. 419: "when asked why he kept silence, he replied 'Why, to let you chatter.'"

41. Ibid., Fr. 165 (Goulet-Cazé trans. IX, 6, p. 1050); Eng. 413.

42. Platon, *Apologie de Socrate*, trans. M. Croiset (Paris: Les Belles Lettres, 1970), 30b, p. 157; English translation by Hugh Tredennick, *Socrates' Defence (Apology)*, in Plato, *The Collected Dialogues*, ed. Edith Hamilton and Huntington Cairns (Princeton, N.J.: Princeton University Press, 1961), 16.

43. On these two notions and their difference, see *L'herméneutique du sujet*, for example, 301–306, 436–437; *The Hermeneutics of the Subject*, 315–321, 454–456.

44. On this aspect of Socratic speech, see the lecture of March 2, 1983, first hour, in *Le gouvernement de soi et des autres*, 286–296; *The Government of Self and Others*, 310–321.

45. Platon, *Apologie de Socrate*, 21a–e, pp. 145–146; Plato, *Socrates' Defence (Apology)*, 7–8.

46. The scene is related by Alcibiades in the *Symposium* (220a–220d). See the reference to this scene in the 1982 lectures, *L'herméneutique du sujet*, 49; *The Hermeneutics of the Subject*, 49–50.

47. See the analysis of this dialogue in the lectures of January 6 and 13, 1982, in *L'herméneutique du sujet*, 3–77; *The Hermeneutics of the Subject*, 1–79.

48. Platon, *Lachès*, trans. M. Croiset (Paris: Les Belles Lettres, 1965), 200e, p. 121; English translation by Benjamin Jowet, *Laches*, in Plato, *The Collected Dialogues*, 144.

ANTHROPOS TODAY: REFLECTIONS ON MODERN EQUIPMENT
PAUL RABINOW

1. Michel Foucault, "Hautes Etudes," in *L'herméneutique du sujet: Cours au Collège de France, 1981–1982*, ed. Frédéric Gros (Paris: Editions de l'Ecole des Hautes Etudes, Editions Gallimard, Editions du Seuil, 2001), 312.

2. Peter Sloterdijk, *Règles pour le parc humain. Une lettre en réponse à la Lettre su l'humanisme de Heidegger*, trans. Olivier Mannoni (Paris: Editions, Mille et Une Nuits, 2000), 7; English trans. Paul Rabinow. Originally published as *Regeln für den Menschenpark: Ein Antwortschreiben zu Heideggers Brief über den Humanismus* (1999).

3. Hubert Dreyfus and Paul Rabinow, *Beyond Structuralism and Hermeneutics*, 2nd ed. (Chicago: University of Chicago Press, 1983).

4. Ibid., xxiii–xxiv.

5. On "pathos," see Paul Rabinow, *French Modern: Norms and Forms of the Social Environment* (Chicago: University of Chicago Press, 1989), 14.

6. Gros, "Situation du cours," Foucault, *L'herméneutique*, 504.

7. Stephen Shapin, in his *Leviathan and the Air-Pump: Hobbes, Boyle, and the Experimental Life* (Princeton, N.J.: Princeton University Press, 1989), [with Simon Shaffer], as well as in personal communication, argues convincingly that the actual history was a good deal more complicated than this version. The moral status of gentlemen scientists, as well as their access to instruments, went a long way to guaranteeing their credibility.

8. Foucault, *L'herméneutique*, 442.

9. Ibid.

10. Ibid., 308.

11. Foucault, "Résumé du cours," 479.

12. Gros, "Situation du cours," 510.

13. Foucault, "Résumé du cours," 479–480.

14. Ibid., 312.

15. Sloterdijk, *Règles pour le parc humain*, 7.
As the lecture was delivered in Basel as part of a cycle of lectures on "the actuality of humanism," its contents are less surprising.

16. Sloterdijk, *Règles pour le parc humain*, 7, my translation.

17. Ibid., 9.

18. On the overemphasis on communication, see Gilles Deleuze and Félix Guattari, *Qu'est-ce que la philosophie?* (Paris: Editions de Minuit, 1991), 12.

19. Sloterdijk, *Règles pour le parc humain*, 18.

20. Ibid., 11: "Les peuples se sont organisés comme *des amicales obligatoires*, intégralement alphabétisées, ne jurant que par un canon de lecture toujours obligatoire dans l'espace national."

21. Ibid., 13.

22. Max Weber, *The Protestant Ethic and the Spirit of Capitalism* (New York: Scribner's, 1950), 182–183.

23. Sloterdijk, *Règles pour le parc humain*, 45.

24. Ibid., 18.

25. Sigmund Freud, *Civilization and Its Discontents*, trans. James Strachey (New York: Norton, 1961), 44–45.

26. Sigmund Freud, "A Difficulty in the Path of Psycho-Analysis," originally printed in *Nyugat* in Hungarian translation, originally translated into English in 1920; Standard Edition, translation, 1925. The "three blows to human narcissism" are also described at the end of lecture 18 of Freud's *Introductory Lectures* (1916–1917), part 3, pp. 284–285.

27. Freud does not discuss resistance to claims put forward by scientists in which resistant opinion turned out to be correct or cases in which science could not provide adequate answers.

28. Immanuel Kant, "What Is Enlightenment?" trans. Louis White Beck, in Immanuel Kant, *On History* (Indianapolis, Ind.: Library of Liberal Arts, 1963).

29. Max Weber, "Science as a Vocation," in H. H. Gerth and C. Wright Mills, eds., *From Max Weber: Essays in Sociology* (Oxford: Oxford University Press, 1946).

30. Weber, *Protestant Ethic*, 182.

31. Weber, "Science as a Vocation," 131.

32. Ibid., 132.

33. Ibid., 134.

34. Ibid.

35. Ibid.

36. Ibid., 137.

37. Ibid., 136.

38. Ibid., 139.

39. Ibid.

40. Ibid., 140.

41. Ibid., 142.

42. Ibid., 143.

43. Ibid., 144.

44. Ibid., 143.

45. Jürgen Habermas, "Two Hundred Years' Hindsight," in James Bohman and Matthias Lutz-Bachmann, eds., *Perpetual Peace: Essays on Kant's Cosmopolitan Ideal* (Cambridge, Mass.: MIT Press, 1997), 124.

46. François Jacob, *Of Flies, Mice, and Men*, trans. Giselle Weiss (Cambridge, Mass.: Harvard University Press, 1998), 96, 92. Originally published as *La souris, la mouche et l'homme* (Paris: Editions Odile Jacob, 1997).

47. On the role of core metaphors, see Hans Blumenberg, *Die Lesbarkeit der Welt* (Frankfurt am Main: Suhrkamp Verlag, 1981).

48. Freud, *Civilization and Its Discontents*, 112.

49. Ibid., 111.

Immanuel Kant, "An Answer to the Question: What Is Enlightenment?" in *Practical Philosophy*, edited and translated by Mary J. Gregor, 11–22. New York: Cambridge University Press, 1999. ["Beantwortung der Frage: Was ist Aufklärung?" *Berlinische Monatsschrift* 4 (1784): 481–494.] Copyright © 1996 Cambridge University Press. Reprinted with the permission of Cambridge University Press.

Max Weber, "Science as a Vocation," in *The Vocation Lectures*, translated by Rodney Livingstone, edited by David Owen and Tracy B. Strong, 1–31. Indianapolis, Ind.: Hackett, 2004. Copyright © 2004 by Hackett Publishing Company, Inc. Reprinted by permission of Hackett Publishing Company, Inc. All rights reserved.

John Dewey, "Reconstruction as Seen Twenty-Five Years Later," in *Reconstruction in Philosophy*, enlarged ed., v–xli. Boston: Beacon, 1948.

Michel Foucault, "What Is Enlightenment?" in *The Foucault Reader*, edited by Paul Rabinow, translated by Catherine Porter, 32–50. New York: Pantheon, 1984. Translation copyright ® 1984 by Random House, Inc. Copyright ® as an unpublished work, 1984, by Michel Foucault and Paul Rabinow. Used by permission of Pantheon Books, an imprint of the Knopf Doubleday Publishing Group, a division of Random House LLC. All rights reserved.

Hans Blumenberg, *The Legitimacy of the Modern Age*, translated by Robert M. Wallace, 229–242, 377–402. Cambridge, Mass.: MIT Press, 1983. © 1983 Massachusetts Institute of Technology, by permission of The MIT Press.

Georges Canguilhem, "The Living and Its Milieu," in *Knowledge of Life*, translated by S. Geroulanos and D. Ginsburg, 98–120. [Original title: "Le vivant et son milieu," in *La connaissance de la vie*, Librairie Philosophique, 129–154. Paris: J. Vrin, 1965.] © Librairie Philosophique, J. Vrin, Paris. http://www.vrin/fr.

Georges Canguilhem, "The Question of Normality in the History of Biological Thought," in *Ideology and Rationality in the History of the Life Sciences*, translated by Arthur Goldhammer, 125–146. Cambridge, Mass.: MIT Press, 1988. [Original title: "La question de la normalité dans l'histoire de la pensée biologique," in *Idéologie et rationalité dans l'histoire des sciences de la vie*, Librairie Philosophique, 121–139. Paris: J. Vrin, 1977.] © Librairie Philosophique, J. Vrin, Paris. http://www.vrin/fr.

Michel Foucault, *The Hermeneutics of the Subject: Lectures at the Collège de France, 1981–1982*, edited by Arnold I. Davidson, translated by Graham Burchell, 1–41. New York: Palgrave Macmillan, 2005. © 2005 by Michel Foucault. Translation © Graham Burchell. Reprinted by permission of Palgrave Macmillan. All rights reserved.

Michel Foucault, *The Courage of Truth—The Government of Self and Others II: Lectures at the Collège de France 1983–1984*, edited by Arnold I. Davidson, translated by Graham Burchell, 1–31. New York: Palgrave Macmillan, 2011. © 2005 by Michel Foucault. Translation © Graham Burchell. Reprinted by permission of Palgrave Macmillan. All rights reserved.

Paul Rabinow, *Anthropos Today: Reflections on Modern Equipment*, 1–12, 76–106. Princeton, N.J.: Princeton University Press, 2003. © 2003 Princeton University Press. Reprinted by permission of Princeton University Press.

forms of living

Stefanos Geroulanos and Todd Meyers, *series editors*

Georges Canguilhem, *Knowledge of Life*. Translated by Stefanos Geroulanos and Daniela Ginsburg, Introduction by Paola Marrati and Todd Meyers.

Henri Atlan, *Selected Writings: On Self-Organization, Philosophy, Bioethics, and Judaism*. Edited and with an Introduction by Stefanos Geroulanos and Todd Meyers.

Catherine Malabou, *The New Wounded: From Neurosis to Brain Damage*. Translated by Steven Miller.

François Delaporte, *Chagas Disease: History of a Continent's Scourge*. Translated by Arthur Goldhammer, Foreword by Todd Meyers.

Jonathan Strauss, *Human Remains: Medicine, Death, and Desire in Nineteenth-Century Paris*.

Georges Canguilhem, *Writings on Medicine*. Translated and with an Introduction by Stefanos Geroulanos and Todd Meyers.

François Delaporte, *Figures of Medicine: Blood, Face Transplants, Parasites*. Translated by Nils F. Schott, Foreword by Christopher Lawrence.

Juan Manuel Garrido, *On Time, Being, and Hunger: Challenging the Traditional Way of Thinking Life*.

Pamela Reynolds, *War in Worcester: Youth and the Apartheid State*.

Vanessa Lemm and Miguel Vatter, eds., *The Government of Life: Foucault, Biopolitics, and Neoliberalism*.

Henning Schmidgen, *The Helmholtz Curves: Tracing Lost Time*. Translated by Nils F. Schott.

Henning Schmidgen, *Bruno Latour in Pieces: An Intellectual Biography*. Translated by Gloria Custance.

Veena Das, *Affliction: Health, Disease, Poverty*.

Kathleen Frederickson, *The Ploy of Instinct: Victorian Sciences of Nature and Sexuality in Liberal Governance*.

Roma Chatterji (ed.), *Wording the World: Veena Das and Scenes of Inheritance*.

Jean-Luc Nancy and Aurélien Barrau, *What's These Worlds Coming To?* Translated by Travis Holloway and Flor Méchain. Foreword by David Pettigrew.

Anthony Stavrianakis, Gaymon Bennett, and Lyle Fearnley (eds.), *Science, Reason, Modernity: Readings for an Anthropology of the Contemporary*.